Unifying Concepts
and
Processes
in
Elementary
Mathematics

University of Maryland
Mathematics Project

ALLYN and BACON, INC.
BOSTON, LONDON, SYDNEY, TORONTO

Library of Congress Cataloging in Publication Data

Maryland.   University.   Mathematics Project.
    Unifying concepts and processes in elementary mathematics

    Includes index.
    1. Mathematics—1961–   I. Title.
QA39.2.M365   1977   510   77–22582
ISBN 0–205–05844–2

Contents

Preface

For more than five years the staff of the University of Maryland Mathematics Project (UMMaP) has been involved in a thorough re-examination of the content, sequence, and pedagogy for a basic course in elementary mathematics. Discussions and classroom experiences led to the following broad principles.

1. A contemporary organization of mathematics must unify the various special topics through the fundamental concepts of sets, operations, relations, mappings, groups, and fields.
2. In any mathematics course there are many important process skills to be acquired along with mathematical facts.
3. It is important that content topics not ordinarily included, such as probability and statistics, computing, and geometric transformations, be represented in an elementary mathematics textbook.
4. Promising pedagogical procedures such as small group learning experiences or classroom discovery techniques should be exploited as effective instructional techniques.

This text, the product of more than four years of development, classroom testing, and evaluation by the entire UMMaP staff, meets such criteria for a sound contemporary program.

The text is particularly well suited for a two semester course in mathematics content for the elementary school teacher, but it is also appropriate as a text for a basic liberal arts course. Often, elementary mathematics courses have emphasized review for mastery of arithmetic, algebra, and geometry. The UMMaP program accomplishes this re-examination while placing each idea in a broader context of mathematical content and process structures. Mathematics is viewed as the study of structure in sets of objects.

The content structures are provided by binary operations, relations, and mappings. Structures are identified, manipulated, and applied by inductive, deductive, and problem-solving processes.

The student text is a sequence of developmental problems and activities designed primarily for use in classrooms organized for small group learning experiences or classrooms organized for discovery lessons led by the instructor. The development of each mathematical topic is initiated by brief exposition, problems, or activities in the text. The problems and activities are elaborated by homework exercises and summary exposition. The students are provided the opportunity to complete the development. Care has been taken to present enough of the development of the mathematics so that beginning college students will be able to pool their resources from past experiences in mathematics to solve the problems. The material was developed in classrooms using small groups and teacher guided discovery to test the material. Most of the material was revised many times based on the reactions of the students. An Instructor's Manual, which accompanies this text, provides suggestions for how to organize the classroom to utilize small group learning and teacher guided discovery approaches to their maximum benefit.

In most cases the homework exercises are for reinforcement of the concepts and processes developed in the sequence of problems. They are generally not integral to the development of the mathematics, although they may involve development of additional concepts. The Instructor's Manual contains answers to the homework exercises and the problems as well as comments and strategies for solving the problems. The Instructor's Manual also contains additional problems for use with students as review questions or test items, exposition about each content and process experience, and estimated class time for each section.

A number of sections in the text have been starred. These are the most likely ones to be omitted as needed. Following are suggestions for courses of different lengths:

Two 4-semester-hour courses:
  1st course   Chapters 1–4
  2nd course   Chapters 5–9

Two 3-semester-hour courses:

    1st course   Chapter 1, All Sections
                    Chapter 2, Sections 2.1–2.10, 2.14, 2.15
                    Chapter 3, Sections 3.1, 3.5–3.7
                    Chapter 4, Sections 4.1–4.11, 4.14

    2nd course   Chapter 5, All Sections
                    Chapter 6, Sections 6.1–6.8
                    Chapter 7, Sections 7.1–7.5
                    Chapter 8, Sections 8.1–8.5
                    Chapter 9, Sections 9.1–9.5

One 3-semester-hour course:

    Chapter 1, All Sections
    Chapter 2, Sections 2.1–2.10, 2.14, 2.15
    Chapter 4, Sections 4.1–4.4, 4.8–4.11, 4.14
    Chapter 5, All Sections
    Additional Selected Topic from Remaining Chapters

A few comments are needed in connection with the computer thread in the course. Both flowcharting and programming in BASIC language are included. The programming instruction in the text is sufficient for the suggested exercises with only minimal material needed to adapt to the individual computer facility. It is not necessary to use the programming in order to have a coherent course; however, it is desirable. More information on computing is contained in the Instructor's Manual.

This text can be the basis of an exciting experience in mathematics. The content is unified around central mathematical concepts, the processes are emphasized for their own value, a wide variety of contemporary topics are included, and the pedagogical possibilities for small group or teacher led discovery learning are provided. The University of Maryland Mathematics Project believes that students will find this integrated experience in mathematics most worthwhile and that instructors will enjoy using the text to help open up the world of mathematics.

Contributing Authors

Sada Chernick
David Cheslock
*Mildred Cole
*Neil Davidson
 Karen Dorsey
*James Fey
 Susan Grant
*James Henkelman
 Richard Hildenbrand
 Bernice Kastner
 Edith Meyers
 David Preston
 Maxine Rosenfeld
 Joan Scott
 Harry Tunis
 Susan Wagner

*Senior staff, University of Maryland Mathematics Project

Acknowledgments

The authors wish to thank Vivianne Merryman, the secretary to The University of Maryland Mathematics Project, for her patience in working with us and her careful attention to the typing of the manuscript.

The authors also thank the following present and past members of the Project for their contributions to the creation of this text: Ann Anderson, Sylvia Auton, Joanne Rossi Becker, Angela Gallichio, Delia Klingbeil, Norman Locksley, William Moulds, Gerald O'Shaughnessy, James Plutschak, Boyce Rogers, and Ward Stewart.

Mathematical Processes

1

1.1 INDUCTION AND DEDUCTION

Mathematics is a two-pronged science. First is the exploratory, the discovery of possible relationships. Second is the confirming, the testing of those possible relationships. In your past work in mathematics, you may have done little of either of these activities—much time was probably spent telling you about a relationship and then having you apply the relationship. That's all right; but the fun of mathematics from kindergarten to graduate school is uncovering ideas and proving them.

Induction and deduction are two words used to describe these processes of exploring and confirming. Perhaps Figure 1.1 will be useful.

The exploring process is very much like exploration in the physical or life sciences. In mathematics the exploration takes place with things like numbers and geometric figures. The confirming process is different. In the other sciences confirmation takes place in the laboratory; in mathematics confirmation is often deductive proof. Deductive proof is the laboratory for mathematics. You'll do both induction and deduction in these materials. First concentrate on the inductive side of mathematics.

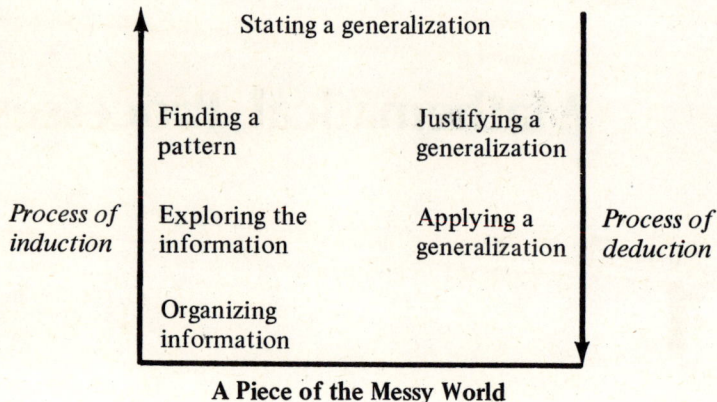

Stating a generalization

Finding a pattern	Justifying a generalization
Exploring the information	Applying a generalization
Organizing information	

Process of induction

Process of deduction

A Piece of the Messy World

FIGURE 1.1

1.2 EXPLORING NUMBER SEQUENCES

PROBLEM 1   Look at the following sequence of numbers: 2, 4, 6, 8, 10, . . .
   a. What do you think will be the next term in the sequence?
   b. Take a look at how you think the sequence is generated, and write what you think the 10th term in the sequence will be.
   c. What will be the 100th term in the sequence?
   d. If we let n represent any counting number, then what will be the nth term in the sequence?
   Note: The set of *counting numbers* is the set of numbers: 1, 2, 3, 4, . . .

For the sequence in Problem 1 you probably came up with the 100th term being 2 × 100 or 200 and the nth term being 2 × n. If you did, you were correct.

PROBLEM 2   Now ask the same questions for the sequence: 1, 3, 5, 7, . . .

PROBLEM 3   Now that you're getting warmed up ask the questions in Problem 1 for sequence: 1, 4, 9, 16, . . .

PROBLEM 4   Try the questions in Problem 1 on the sequence: 1, 4, 7, 10, . . .

PROBLEM 5   Now it's time for one which seems a little bit more complicated. Ask the same questions as in Problem 1 for the sequence: 2, 6, 12, 20, . . .

Were you able to find the *n*th term in Problem 5? If you couldn't, check with your instructor. Only an unusual group will find the *n*th term in Problem 5. There's nothing wrong with not being able to solve all problems. However, don't give up too quickly; sometimes you'll surprise yourself.

PROBLEM 6   Work on the following sequence: 2, 9, 28, ... Ask the same question you asked in Problem 1.

You may have gotten into an argument over the sequence in Problem 6. What was the 4th term in the sequence? Did you get 65? Or did you get 59? Or did you get something else? There is a good explanation for arriving at either 65 or 59. The *n*th term might be $n^3 + 1$ or it might be $(6 \times n^2) - (11 \times n) + 7$. Of course you might have a good explanation for still a different pattern. There is *not* always *one* right answer when you are pattern hunting.

PROBLEM 7   Try asking the same questions as in Problem 1 for each of the following four sequences:

    4, 8, 12, ...
    5, 12, 19, ...
    5, 13, 23, ...
    7, 12, 17, ...

Check your answers with your instructor.

HOMEWORK EXERCISES

You've been organizing and exploring the information, finding a pattern, and stating a generalization. Now invent a few sequences for others to explore.

1.3 CONFIRMING TAKE-A-NUMBER PUZZLES

PROBLEM 1   Just follow the directions. Each one in your group should follow the directions at the same time using his or her own counting number. Don't let the others know what number you started with.

TAKE ANY COUNTING NUMBER.
ADD THREE.
DOUBLE THE RESULT.
SUBTRACT FOUR.
DIVIDE BY TWO.

If you told me your final result, I could tell you what number you started with. Compare the starting numbers of each person in the group with his or her final results, and see if you can figure out the pattern between a starting number and the result.

In Problem 1 you probably found that the result is one more than the starting number. One way to state the generalization for this pattern is: If n is the starting number, then $n + 1$ is the resulting number.

PROBLEM 2   Try analyzing why this take-a-number puzzle works this way by filling in Table 1.1. Have each person in your group fill in a line starting with the number he or she picked in the first column. Put the appropriate results in each of the remaining columns.

TABLE 1.1

Take a number	Add three	Double the result	Subtract four	Divide by two

It is easy to see by looking at the starting number and final result that the result is one more than the starting number.

One way to confirm that the result in Problem 1 will always be one more than the starting number is to test some more examples as you did in Table 1.1. This kind of confirmation is valuable. However, there is always the chance that you will find a different result for some counting number.

PROBLEM 3   Try several more examples.

TABLE 1.2

Take a number	Add three	Double the result	Subtract four	Divide by two

Did you find any results that were different from your generalization?

This process of testing the generalization by trying a number of examples is much like what the scientist does when confirming theories in the laboratory. Contradictory results may be found and then the theory must be modified. But confirmation in mathematics can be more powerful. You can test the result in the set of all counting numbers. The way to begin that process is to express the result you have generalized for all counting numbers.

PROBLEM 4   Fill in Table 1.3 using n to represent the counting number you started with.

TABLE 1.3

Take a number	Add three	Double the result	Subtract four	Divide by two

Your row should have read n, $n + 3$, $2 \times (n + 3)$ or $2n + 6$, $2 \times n + 2$, $n + 1$. As you know, the $\times$, the $\cdot$, or no symbol at all can represent multiplication. Use whichever one you find most convenient: for example, $2 \times n + 2$ or $2 \cdot n + 2$ or $2n + 2$. Now if each of these steps in the last row can be justified in terms of the way all counting numbers work, then the confirmation is a rather good one mathematically. Discuss this and then compare your justification with the following:

Take a number	n	
Add three	$n + 3$	
Double the result	$2(n + 3)$ or $2n + 6$	One of the properties of counting numbers justifies why you rewrite $2(n + 3)$ as $2n + 6$
Subtract four	$2n + 2$ or $2(n + 1)$	The same property justifies rewriting $2n + 2$ as $2(n + 1)$
Divide by two	$n + 1$	

Your justification may be more complete or less complete than this. In this course you will use the deductive process often to justify generalizations in different settings.

PROBLEM 5   Try the following take-a-number puzzle.

TAKE ANY COUNTING NUMBER.
MULTIPLY THAT COUNTING NUMBER BY ITSELF.
SUBTRACT THE NUMBER YOU STARTED WITH FROM THE RESULT.
DIVIDE THE RESULT BY THE NUMBER YOU STARTED WITH.
ADD ONE.

Find the pattern between the starting number and the final result. After you have found the pattern express the result as a generalization. Now confirm this generalization using what you know about counting numbers.

The pattern in Problem 5 was an easy one since your final result was the same as the number you started with. There is one interesting aspect to this one. It is important that you start with a counting number, because if you started with 0, you would run into difficulty. Try starting with 0 and see if you find the difficulty before reading further.

The difficulty in starting with 0 in Problem 5 comes in the fourth step where you divide by the number you started with. You started with 0 and division by 0 is not defined. You will run into this division by 0 situation many more times.

PROBLEM 6   Try still another take-a-number puzzle.

TAKE A COUNTING NUMBER.
ADD TWO.
TRIPLE THE RESULT.
SUBTRACT THE NUMBER YOU STARTED WITH.
DIVIDE BY TWO.
SUBTRACT THE NUMBER YOU STARTED WITH.

First try this take-a-number puzzle in your group and look for a pattern in the final results. Now confirm this generalization.

You have now had some experience with the deductive side of mathematics in your confirming activities.

HOMEWORK EXERCISES

1. The results of the following take-a-number puzzle may be surprising.
   a. TAKE A COUNTING NUMBER.
      MULTIPLY THE COUNTING NUMBER BY ITSELF.
      SUBTRACT THE NUMBER YOU STARTED WITH.
      ADD ONE.
      SUBTRACT THE NUMBER YOU STARTED WITH.
      ADD ONE.
      SUBTRACT THE NUMBER YOU STARTED WITH.
   b. One person tried 1 and got a final result of 0.
      Another person tried 2 and got a final result of 0.
      Test these results for yourself.
      Are you ready to claim a pattern in the results?
   c. Test that pattern to be sure it is true no matter what number you start with.

2. Write a couple of your own take-a-number puzzles.
   a. Write one that will give you a result which is double the starting number.
   b. Write one that will always give you a result of two.
   c. Write one that will do whatever you want it to do.
   d. Try them out on someone.

3. You have generated some formulas and tested them in these take-a-number puzzles. Now try testing out some formulas others have put

together. Some of them are formulas which are always true, others are formulas which never are true, and some are true most of the time, but not all of the time. Assume that all of the formulas are about the counting numbers.

For each of the following, decide if the formula is

a. true for all counting numbers,
b. false for all counting numbers, or
c. true only for certain counting numbers.
   List the exceptions.
a. Check the formula $n < n + 1$. ($<$ means "is less than")
b. Check the formula $2^n > 2n + 1$ as above. ($>$ means "is greater than")
c. Check the formula $3 \cdot (n + 2) = (3 \cdot n) + (3 \cdot 2)$ as above.
d. Check the formula $3 + (n \cdot 2) = (3 + n) \cdot (3 + 2)$ as above.
e. Check the formula $\dfrac{n^2 - 1}{n + 1} = n - 1$ as above.
f. Finally check the formula $\dfrac{n^2 - 3n + 2}{n - 2} = n - 1$ as above.

1.4 PROBLEM SOLVING

You have done some induction and you have done some deduction by finding patterns and testing them. These processes interact with each other in many ways. In fact, they are certainly a part of problem solving. But there are other parts of problem solving. One aspect of problem solving is making sure the information is clear. For instance, consider the problem that follows:

Problem: In an *n*-by-*n* square array of dots, how many dots are there in the array that are *not* on the main diagonal?

There is a need for clear communication. What is a square array of dots? What is the main diagonal? Is *n* a counting number?

Let us assume that *n* is a counting number. The main diagonal starts at the upper left-hand corner as in Figure 1.2, an example of a 5-by-5 array of dots.

Just considering a 5-by-5 array of dots is a start toward solving the problem. It is a specific case and a relatively simple

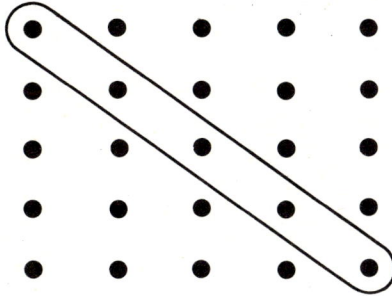

FIGURE 1.2

one. In fact, it is easy enough to count the dots that are not on the main diagonal. There are 20.

An even simpler example would be a 3-by-3 array of dots. If you draw the 3-by-3 array, there are 6 dots in the array that are not on the main diagonal.

Now is the time for another aspect of problem solving—organizing the information or data. Sometimes a table is helpful.

TABLE 1.4

Number of Dots on Side of Array	3	5			n
Number of Dots Not on Main Diagonal	6	20			?

You might try to fill in other possibilities. Also, in this case, it turns out that adding a little more information to the table helps clarify the pattern.

TABLE 1.5

Number of Dots on Side of Array	3	5			n
Total Number of Dots in Array	9	25			?
Number of Dots Not on Main Diagonal	6	20			?

Now finish the Problem.

1.5 PROBLEMS

Use all of your problem solving ideas including full use of induction and deduction as you work on some of the following varied problems.

PROBLEM 1   *Diagonals of a Polygon*
Find the number of diagonals in a regular polygon of *n* sides where *n* is a counting number.

PROBLEM 2   *Abracadabra†*
The word *abracadabra* means something like "complicated nonsense." The word is used contemptuously today, but there was a time when it was a magic word, engraved on amulets in mysterious forms (see Figure 1.3), and people believed that such an amulet would protect the wearer from disease and bad luck.

```
                    A
                 B     B
              R     R     R
           A     A     A     A
        C     C     C     C     C
     A     A     A     A     A     A
        D     D     D     D     D
           A     A     A     A
              B     B     B
                 R     R
                    A
```

FIGURE 1.3

In how many ways can you read the word *abracadabra* in Figure 1.3? It is understood that you begin with the uppermost *A* (the north corner) and read down, passing each time to the next letter (southeast or southwest) till you reach the last *A* (the south corner).

† Problem 2 adapted from George Polya, *Mathematical Discovery, Volume 1* (New York: John Wiley and Sons, 1962), p. 68.

PROBLEM 3   *The Fraternity House Problem*

Two friends from college met having not seen each other for many years. Joe Shlobotnik says to Ron Roma, "How are you? Did you ever marry Rhona?"

"I'm fine!" says Ron, "and Rhona Klinkdorf is now Rhona Roma."

"Gee, that's great!" says Joe. "Do you have any kids?"

"I have three children," says Ron.

"What are their ages?" queries Joe.

Remembering that Joe likes riddles, Ron replies, "The product of their ages is 36."

"But that doesn't tell me how old your children are," complains Joe.

"O.K.," says Ron. "The sum of their ages is equal to our old fraternity house number."

"Yes, I remember the old frat house number, but I still don't know how old your children are," says Joe.

"Well, my oldest one looks like me," beams Ron proudly.

Joe thinks for a moment and then exclaims, "Now I know how old your children are!"

How old are Ron's 3 children?

PROBLEM 4   *Synthetic Vitamins*

A chemist working for a large chemical company wishes to manufacture a synthetic vitamin pill. She needs to know the order of the ten amino acids that make up the structure of the natural vitamin.

The following notation is adopted. The ten amino acids are numbered 0 through 9. The symbol 3(245) means that 3 is the first amino acid in a fragment chain and the order of the others in parentheses is unknown. The task is to determine the order of the amino acids in each fragment and then to combine fragments to reconstruct their exact order in the original vitamin. The vitamin is 0(01234556789), which means that the amino acid numbered 0 comes first in the chain and the order of the others is unknown. It is known that there are two of the amino acids numbered 5 and two numbered 0. The partially known structures of the 13 fragments follow:

I	0(2345)
II	0(19)
III	0(16789)
IV	1(68)

V	2(55)
VI	2(0559)
VII	3(245)
VIII	4(25)
IX	4(012559)
X	5(019)
XI	5(01569)
XII	6(78)
XIII	9(168)

Use the information gained from the fragments to find the exact order of the amino acids in the original vitamin.

PROBLEM 5   *The Logical Lie Detector*†

Sir Arthur Conan Doyle was a master at weaving tangled tales of intrigue, whose plots were eventually unraveled by the bold logic of Sherlock Holmes. "Elementary, my dear Watson," says Sherlock, and so it is, if you can keep track of the numerous clues leading to the one inescapable conclusion. If you feel up to tackling this type of mystery, join some mathematical detectives and take a crack at the task of tracking down a murderer. The story goes as follows:

Oliver Laurel was killed on a lonely road two miles from Trenton at 3:30 A.M., February 14. By Washington's Birthday, the police had rounded up five suspects in Philadelphia: in alphabetical order, Hank, Joey, Red, Shorty and Tony. Under questioning, each of the men made four statements, three of which were true and one of which was false. It is known that one of these men killed Oliver Laurel (we have that assurance from Charlie the Canary), and the task is to identify the murderer. From the police records, we obtain the following transcript of the statements:

Hank:

a. I did not kill Laurel.

b. I never owned a revolver in my life.

c. Red knows me.

d. I was in Philadelphia the night of February 14.

Joey:

a. I did not kill Laurel.

b. Red has never been in Trenton.

c. I never saw Shorty before.

d. Hank was in Philadelphia with me the night of February 14.

---

† Problem 5 reprinted with permission from Philip J. Davis and William G. Chinn, *3.1416 and All That* (New York: Simon and Schuster, 1969), pp. 33–34.

Red:

a. I did not kill Laurel.

b. I have never been in Trenton.

c. I never saw Hank before now.

d. Shorty lied when he said I am guilty.

Shorty:

a. I was in Mexico City when Laurel was murdered.

b. I never killed anyone.

c. Red is the guilty man.

d. Joey and I are friends.

Tony:

a. Hank lied when he said he never owned a revolver.

b. The murder was committed on St. Valentine's Day.

c. Shorty was in Mexico City at that time.

d. One of us is guilty.

1.6 FLOWCHARTING

Many tasks and procedures in mathematics (and in other areas) can be broken down into a series of steps to help understand the task or procedure more clearly. For example, the take-a-number puzzle in Problem 4 of Section 1.3 is repeated:

1. Take a number n

2. Add three $n + 3$

3. Double the result $2(n + 3)$ or $2n + 6$

4. Subtract four $2 \cdot (n + 3) - 4$ or $2n + 2$

5. Divide by two $(2 \cdot (n + 3) - 4) \div 2$ or $n + 1$

Notice that in this problem, the order in which the steps are performed is very important. If Steps 2 and 4 are interchanged, the result is entirely different. One widely used method of describing sequential procedures of this type is *flowcharting*. Flowcharts are diagrams that indicate which steps are to be performed and the order in which they are to be performed. Figure 1.4 is the flowchart for the take-a-number puzzle.

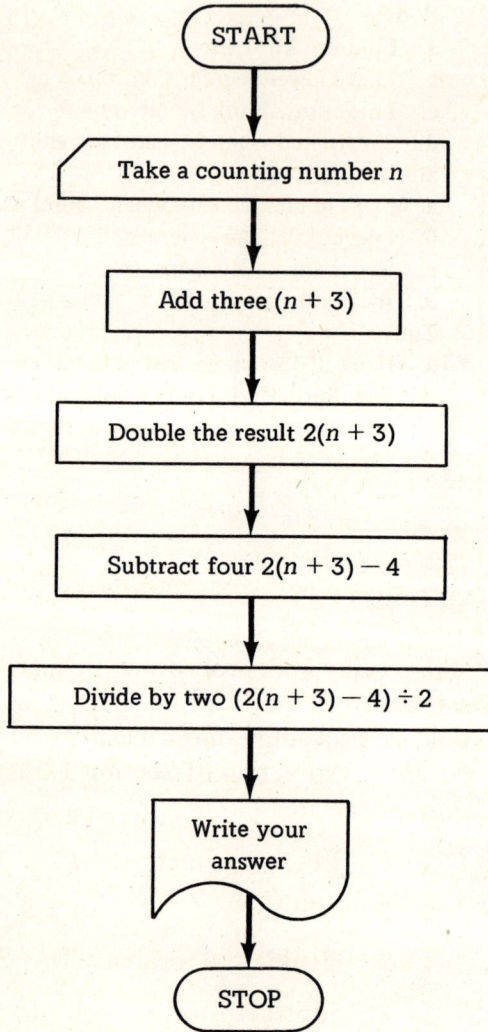

START

Take a counting number n

Add three $(n + 3)$

Double the result $2(n + 3)$

Subtract four $2(n + 3) - 4$

Divide by two $(2(n + 3) - 4) \div 2$

Write your answer

STOP

FIGURE 1.4

Note: Once you begin the flowchart at START, the arrows tell you which step to perform next; there is never a doubt as to the order of the steps. Every rectangle has an arrow coming into it and one leaving it. A diamond can have more than one arrow leaving it.

The shapes of the boxes in a flowchart are not critically important. However, to help readers identify key phases of the procedure, the following notational conventions have developed.

1. Start or Stop	circles or ovals	
2. Input data	the shape of a computer card	
3. Commands or instructions	rectangular boxes	
4. Decision points	diamonds	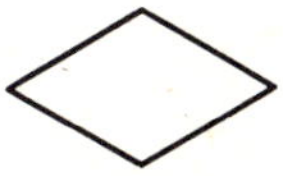
5. Output of results	shapes that suggest a piece of torn paper	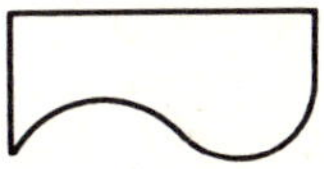

FIGURE 1.5

PROBLEM 1   Construct a flowchart for the following take-a-number puzzle:

TAKE A COUNTING NUMBER.
MULTIPLY THAT NUMBER BY ITSELF.
SUBTRACT THE NUMBER YOU STARTED WITH.
DIVIDE THIS RESULT BY THE NUMBER YOU STARTED WITH.
ADD ONE.

PROBLEM 2   To illustrate the use of the decision box in Figure 1.5, consider Figure 1.6, a flowchart for frying a sunny-side-up egg.
    a.   Fill in the diamond with an appropriate yes or no question.
    b.   Why are there 2 arrows coming from the diamond? Does this cause uncertainty as to which step is to be performed next?
    c.   Describe the part of the flowchart which could be labeled a *loop*.
    d.   What is the purpose of this loop?

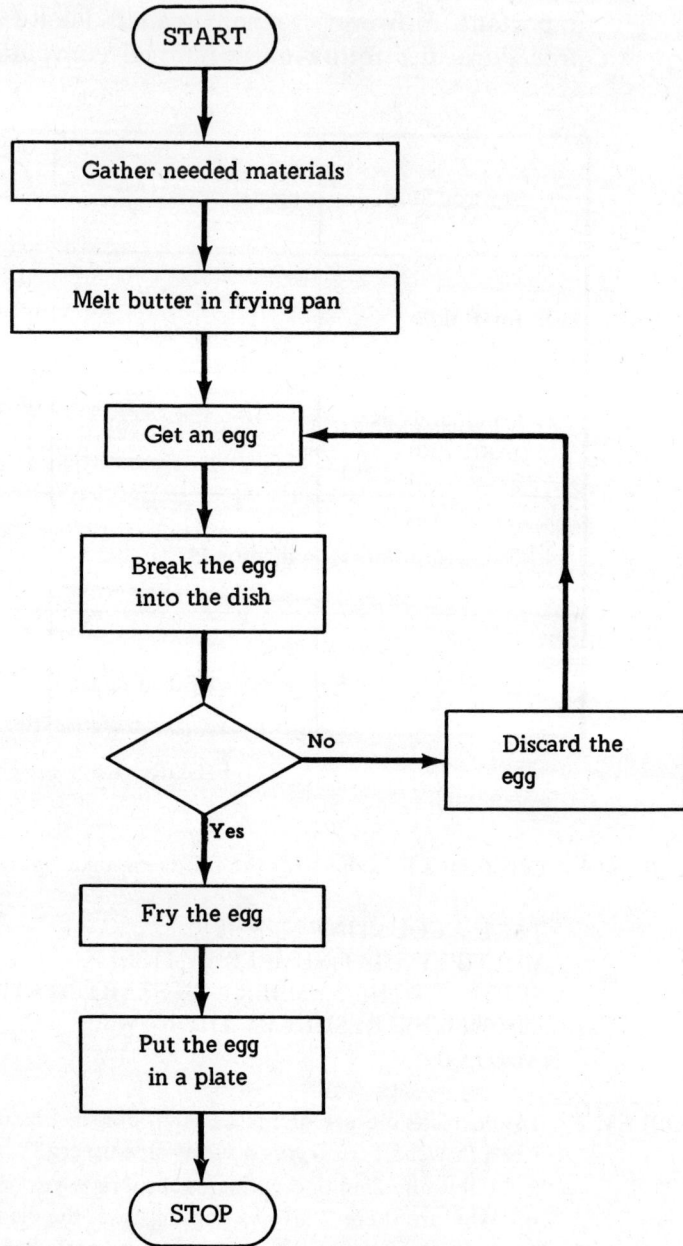

FIGURE 1.6

PROBLEM 3   Recall the take-a-number puzzle in Problem 5 of Section 1.3. You drew a flowchart for this puzzle in Problem 1 of this section.

    a. If step one of this puzzle is changed from TAKE ANY *COUNTING* NUMBER, to TAKE ANY NUMBER, then a diamond-shaped box must be added to the flowchart. Why?

    b. Why is it necessary that the diamond be inserted in the flowchart before the step, DIVIDE BY THE NUMBER YOU STARTED WITH?

    c. Assume that step one has been changed to TAKE ANY NUMBER. Redo your flowchart including the diamond-shaped decision box. **Note:** The only type of question allowed in a decision box is that which can be answered yes or no.

Flowcharting is a process which forces you to analyze a problem and break it down into an ordered sequence of steps. Essentially, it is a way of organizing your thoughts about a particular problem and representing this organization with a diagram. This method is used in business, industry, education, and most extensively in the growing fields of computer science and data processing.

HOMEWORK EXERCISES

Construct flowcharts for each of the following tasks:

1. TAKE A COUNTING NUMBER.
   DOUBLE IT.
   TRIPLE THE RESULT.
   SUBTRACT FIVE.
   ADD TWO.
   DIVIDE BY THREE.
   ADD ONE.

2. TAKE ANY NUMBER.
   ADD FIVE.
   MULTIPLY BY THE ORIGINAL NUMBER.
   ADD FOUR.
   DIVIDE BY ONE MORE THAN THE ORIGINAL NUMBER.
   SUBTRACT THE ORIGINAL NUMBER.

3. Putting on your shoes and socks. (Allow for possibility of laces.)

4. Opening a combination lock.

5. Mixing green paint from blue and yellow.

SUMMARY

Chapter 1 has been a brief look at some of the mathematical processes. You used several inductive processes:

> organizing data to find a pattern;
> searching for a pattern;
> stating generalizations based on patterns in data.

You also used at least a couple of deductive approaches:

> testing a generalization with an example;
> justifying a generalization.

In addition you have used problem solving processes:

> clarifying information and conditions;
> selecting data which is relevant;
> using diagrams, charts, and tables to organize data;
> describing sequential steps in a flowchart;
> reducing a problem to a simpler case.

These processes are of greater importance than specific results. Often in school the only interest in mathematics is in one correct answer. Mathematics is much more than that! Often there is more than one answer; sometimes there is no definite answer. The way in which you think in the subject is important so that you will be able to use mathematics in new situations. Watch the way you approach each problem as you proceed through this text. Take the time to explore in some depth. Use as many of the processes as you can. Exploring a concept, seeing new relationships between concepts, and coming to new understandings will provide as much or more satisfaction as finding a correct answer—if there is one.

Structure in Number Systems

2

2.1 INTRODUCTORY CALENDAR PROBLEMS

The power of mathematics is demonstrated most clearly when careful and clever thinking solves a problem that seemed to promise long and tedious work. Each of the following problems can be solved with patience and a lot of arithmetic calculation. But with some thought and logical reasoning, each will yield to the same type of shortcut strategy.

PROBLEM 1 What day of the week will it be
    a. 6 weeks from today?
    b. 10 weeks and 4 days from today?
    c. 122 days from today?
    d. exactly one year from today?
    e. exactly two months from today?

PROBLEM 2 John F. Kennedy was President of the United States for 1036 days until his death on Friday, November 22, 1963. On what day of the week was he inaugurated?

PROBLEM 3 A luxurious round-the-world cruise departs from Norfolk on Friday. Scheduled sailing time to Rio is 11 days; from Rio arrival to Capetown is 9 days; from Capetown to Singapore is 17 days. On what day of the week will the ship arrive in Rio? In Capetown? In Singapore?

PROBLEM 4   If January 1, 1900, was Monday, on what day of the week will the year 2000 begin?

2.2 OPERATION TABLES FOR Z_7

The idea that simplifies each of the calendar questions is basic to a new system of arithmetic, *modulo seven* (mod 7). The following mod 7 addition table has been started.

PROBLEM 1   Figure out how the given table entries were calculated. Then complete the table.
Note: The table is to be read with the first number read from the left side and the second number read from the top. For example, the circled 4 is the answer to 5 + 6.

TABLE 2.1   Mod 7 Addition

+	0	1	2	3	4	5	6
0	0	1		3			6
1	1	2					0
2	2			5			1
3							
4			6			2	
5							④
6				3			

PROBLEM 2   Using the mod 7 addition table, complete each of the following open sentences correctly.
a.   4 + 5 = _____
b.   6 + 4 = _____
c.   2 + 0 = _____
d.   5 + 3 = _____
e.   3 + 5 = _____
f.   5 + _____ = 4
g.   3 + _____ = 5
h.   _____ + 3 = 1
i.   _____ + 6 = 0

j. $4 + (3 + 6) =$ _____

k. $(4 + 3) + 6 =$ _____

PROBLEM 3   As a check on your arithmetic understanding, complete the following open sentences without using the table.

   a. $3 + 4 =$ _____

   b. $5 + 5 =$ _____

   c.   _____ $+ 3 = 2$

Arithmetic modulo seven involves the numbers 0 thru 6 and new kinds of arithmetic operations. In reference to this system, the set of numbers is usually named $Z_7 = \{0, 1, 2, 3, 4, 5, 6\}$.

The notation $(Z_7, +)$ refers to the set Z_7 together with the operation of addition modulo seven. $(Z_7, +)$ is an example of a *mathematical system,* i.e., a set of numbers and an operation defined on that set.

$(W, +)$ is another common mathematical system. Here, W is the set of all *whole numbers,* i.e., $W = \{0, 1, 2, 3, 4, \ldots\}$, and the operation is ordinary addition.

Note: the same symbol, $+$, is used in $(Z_7, +)$ and $(W, +)$ but with different meanings. The context will always indicate when to add in mod 7 and when to add in ordinary whole number arithmetic.

PROBLEM 4   The tables for mod 7 addition and ordinary whole number addition have many similarities and many striking differences. Study Tables 2.2 and 2.3. Then list

   a.   Some of the patterns in $(Z_7, +)$.

   b.   Some of the contrasts between $(Z_7, +)$ and $(W, +)$.

TABLE 2.2   Mod 7 Addition

+	0	1	2	3	4	5	6
0	0	1	2	3	4	5	6
1	1	2	3	4	5	6	0
2	2	3	4	5	6	0	1
3	3	4	5	6	0	1	2
4	4	5	6	0	1	2	3
5	5	6	0	1	2	3	4
6	6	0	1	2	3	4	5

TABLE 2.3 Whole Number Addition

+	0	1	2	3	4	5	6	7 . . .
0	0	1	2	3	4	5	6	
1	1	2	3	4	5	6	7	
2	2	3	4	5	6	7	8	
3	3	4	5	6	7	8	9	
4	4	5	6	7	8	9	10	
5	5	6	7	8	9	10	11	
6	6	7	8	9	10	11	12	
7								
.								
.								
.								

HOMEWORK EXERCISES

1. Using the mod 7 addition table, complete each of the following open sentences correctly.

   a. $6 + 0 = $ _____

   b. $0 + 4 = $ _____

   c. $4 + $ _____ $ = 0$

   d. $3 + $ _____ $ = 3$

   e. $(6 + $ _____ $) + 2 = 3$

   f. $6 + ($ _____ $ + 2) = 3$

2. In each of the following, find values (if possible) of a, b, and/or c in $(Z_7, +)$ that make the given sentences true statements.

   a. $a + 5 = 5$

   b. $a + a = 3$

   c. $a + b = a$

   d. $3 + c = 4$

   e. $3 + c = 2$

   f. $a + b = a + c$

3. a. The same idea used to produce the $(Z_7, +)$ table can be extended to multiplication. Study the entries already given in the mod 7 multiplication table that follows, then complete the table.

TABLE 2.4   Mod 7 Multiplication

·	0	1	2	3	4	5	6
0		0	0	0	0		
1							
2		2				1	3
3					5		
4	0			1			3
5						6	4
6		6			4		

b. Use the above table to complete correctly each of the following open sentences.

$5 \cdot 1 =$ _____   _____ $\cdot 4 = 4$ _____ $\cdot 6 = 0$ $3 \cdot 5 =$ _____

$5 \cdot 3 =$ _____ $4 \cdot$ _____ $= 1$ _____ $\cdot 5 = 1$

2.3   STATEMENTS ABOUT $(Z_7, +)$

Throughout this section you will be given a number of statements about the system $(Z_7, +)$. You will be asked to make educated guesses about which statements are true and which are false. In making your decision about a given statement, the following advice may be helpful.

Many of the statements below contain the phrase, "for all a, b, in Z_7." To show that such a statement is false, it is sufficient to find one pair of elements a, b in Z_7 for which the statement is false.

Example
$a + b = 0$ for all a, b in Z_7.

This statement is false because $1 + 2 \neq 0$. The symbol, $\neq$, is read is not equal to.

On the other hand, to show the truth of a statement with phrase, "for all a, b, in Z_7," it is *not* sufficient to consider only one pair of elements for which the statement is true. If you wish to prove a statement by considering specific pairs of elements, you need to show that the statement is true for all possible pairs. For some statements, it is an easy task to check all cases, that is all possible pairs.

This testing of all possibilities constitutes a *proof by exhaustion* because you are exhausting all possibilities. Some statements lend themselves to this type of proof, e.g., Problem 1, Statement **f**. When you have confirmed your educated guess about a question by checking all cases, you have actually proven the statement.

For many of the following statements which seem true, a "proof by exhaustion" would be a tedious task. For example, Problem 1, Statement **j** would require the testing of 343 cases. Before making your educated guess about such statements, you should try several examples. Note that trying several examples does *not* constitute a proof.

One last comment before you get started. Consider the following statement: $a + b = a$ for all a, b in Z_7. In this statement, a *must* represent the same value on both sides of the equation. On the other hand, b may or may not represent the same value as a.

PROBLEM 1   Make an educated guess about the truth or falsity of each statement in the following list, and mark T or F in the space provided. If you actually *prove* that a statement is true by considering all possibilities, write *proven*.

a. _____ $a + a = a$ for all a in Z_7.

b. _____ There is an element x in Z_7 such that $a + x = a$ for all a in Z_7. (This element x is called the *identity element* for $(Z_7, +)$.)

c. _____ For each a in Z_7, there is an element x in Z_7 such that $a + x = 0$. (**Note:** If $a + x = 0$, we write $x = {}^-a$ and read x is the *additive inverse* of a.)

For your convenience, complete the following additive inverses in Z_7.

$${}^-0 =$$
$${}^-1 =$$

$$^-2 =$$
$$^-3 =$$
$$^-4 =$$
$$^-5 =$$
$$^-6 =$$

d. _____ There are two different elements x and y in Z_7, such that $3 + x = 0$ and $3 + y = 0$.

e. _____ For every two elements a and b in Z_7, $a + b$ is an element of Z_7.

f. _____ $^-(^-a) = a$ for all a in Z_7.

g. _____ $a + b = b + a$ for all elements a and b in Z_7.

h. _____ $a + (^-b) = b + (^-a)$ for all a, b in Z_7. (Remember that a and b *may* represent the same value. For example, first try $a = 3$, $b = 3$ and then try $a = 2$, $b = 4$.)

i. _____ $^-(a + b) = {^-a} + {^-b}$ for all a, b in Z_7.

j. _____ $a + (b + c) = (a + b) + c$ for all elements a, b, c in Z_7. (This problem can be checked on a computer.)

k. _____ $(a + b) + c = a + (b + c)$ for all elements a, b, c, in Z_7.

Think a moment about the kind of activity that you have done up to this point. You have considered a finite set of numbers, the set Z_7, and explored how the operation, addition, organizes this set of numbers. For example, you've developed the tables for $(Z_7, +)$ to see how addition acts on any two elements of Z_7. You have conjectured about the truth or falsity of statements involving addition in Z_7 and even tested or proved some of these conjectures. Some elements have taken on special roles. For example, the element 0 became an identity and *4* became the inverse of *3*. These roles are dependent on the operation addition. Soon you'll see how these roles change when a new operation, multiplication, is considered on Z_7.

HOMEWORK EXERCISES

In Exercises 1 to 4, decide T or F. Exercises preceded by a star are considered more challenging.

1. _____ If $a = {^-b}$ then $b = {^-a}$.

2. _____ $a + (b + c) = (c + b) + a$ for all $a, b, c,$ in Z_7.

3. _____ If $a + b = c,$ then $b + c = a$ for all a, b, c in Z_7.

4. _____ There is an element a in Z_7 such that $1 + a = a$.

5. Find all elements a in Z_7 such that $a + a = 0$.

★6. Find the smallest counting number n such that $\underbrace{a + a + \cdots + a}_{n \text{ terms}} = 0$ for all a in Z_7.

The three dots in $a + a + \cdots + a$ are included to indicate that a variable number of as are possible; in fact there are n a's.

2.4   MORE ON THE MULTIPLICATION TABLE FOR Z_7

PROBLEM 1   You have completed the table for $(Z_7, \cdot)$ as homework. Use your table to complete correctly each of the following open sentences.
   a. $3 \cdot (2 \cdot 5) = $ _____
   b. $(3 \cdot 2) \cdot 5 = $ _____
   c. $(4 \cdot $ _____ $) \cdot 6 = 3$
   d. $4 \cdot ($ _____ $\cdot 6) = 3$
   e. $(^-3) \cdot (^-2) = $
   Note: This means the additive inverse of 3 multiplied by the additive inverse of 2.
   f. $3 \cdot 2 = $ _____
   g. $4 \cdot ($ _____ $+ 5) = 4 \cdot 6 + 4 \cdot 5$

PROBLEM 2   As with addition, mod 7 multiplication and ordinary whole number multiplication have many similarities and differences. Study Tables 2.5 and 2.6. Then list
   a. Some of the patterns in $(Z_7, \cdot)$
   b. Some of the contrasts between $(Z_7, \cdot)$ and $(W, \cdot)$

PROBLEM 3   In each of the following, find values of $a, b,$ and/or c in Z_7 that make the given sentences true statements.
   a. $a \cdot b = a$
   b. $c \cdot c = 1$
   c. $^-a \cdot b = a \cdot {}^-b$
   d. $a \cdot b = a \cdot c$
   e. $a \cdot a = a$

TABLE 2.5   Mod 7 Multiplication $(Z_7, \cdot)$

·	0	1	2	3	4	5	6
0	0	0	0	0	0	0	0
1	0	1	2	3	4	5	6
2	0	2	4	6	1	3	5
3	0	3	6	2	5	1	4
4	0	4	1	5	2	6	3
5	0	5	3	1	6	4	2
6	0	6	5	4	3	2	1

TABLE 2.6   Whole Number Multiplication $(W, \cdot)$

·	0	1	2	3	4	5	6	7 ...
0	0	0	0	0	0	0	0	
1	0	1	2	3	4	5	6	
2	0	2	4	6	8	10	12	
3	0	3	6	9	12	15	18	
4	0	4	8	12	16	20	24	
5	0	5	10	15	20	25	30	
6	0	6	12	18	24	30	36	
7								
⋮								

2.5 STATEMENTS ABOUT $(Z_7, \cdot)$

PROBLEM 1   Decide whether each statement in the following list is true or false, and mark T or F in the space provided. If you actually *prove* that a statement is true by considering all possibilities, write *proven.*

a. _____ There is an element x in Z_7 such that $a \cdot x = a$ for all a in Z_7. (This element x is called the identity element for $(Z_7, \cdot)$.)

b. _____ For any elements a, b in Z_7, $a \cdot b$ is an element of Z_7.

c. _____ There is an x in Z_7 such that $x \cdot 0 = 1$.

d. _____ For each $a \neq 0$ in Z_7, there is an element x in Z_7 such that $a \cdot x = 1$.

(**Note:** If $a \cdot x = 1$, we write $x = a^{-1}$ and read x is the _multiplicative inverse_ of a.)

For your convenience, complete the following table of multiplicative inverses in Z_7.

$$1^{-1} = \qquad 3^{-1} = \qquad 5^{-1} =$$
$$2^{-1} = \qquad 4^{-1} = \qquad 6^{-1} =$$

e. _____ $a \cdot b^{-1} = b \cdot a^{-1}$ for all $a, b \neq 0$ in Z_7

f. _____ $a \cdot b = b \cdot a$ for all a, b in Z_7

g. _____ $(a \cdot b) \cdot c = a \cdot (b \cdot c)$ for all a, b, c in Z_7. (This problem can be checked by the computer.)

h. _____ $a^{-1} \cdot b^{-1} = (a \cdot b)^{-1}$ for all permissible elements a, b in Z_7. (For this statement to make sense, which elements must be nonzero?)

i. _____ There are more than two elements x in Z_7 such that $x \cdot x = x$.

PROBLEM 2   If possible, complete the following statement so that it is true for all a in Z_7: $(a^{-1})^{-1} =$ _____.

Discuss any difference in your answer if you complete the statement so that it is true for all $a \neq 0$ in Z_7.

At this point you've considered how a new operation, multiplication, has organized the set Z_7. Your investigation has been similar to the case of addition in Z_7. The structure is different though. For example, 0 no longer plays the role of identity, as in addition, but takes on another role: $a \cdot 0 = 0$ for all a in Z_7. The element 1 becomes the identity, and whereas 4 was the inverse of 3 in addition, 5 takes this role in multiplication: $5 \cdot 3 = 1$.

HOMEWORK EXERCISES

In exercises 1 to 7, decide T or F.

1. _____ $a \cdot b^2 = a^2 \cdot b$ for all a, b in Z_7. (**Note:** $a^2 = a \cdot a$)

2. _____ For all a, b, c in Z_7, if $a \cdot b = a \cdot c$ and $a \neq 0$, then $b = c$.

3._____ $a \cdot b = a^2 \cdot b^2$ for some a, b in Z_7 such that $a \neq b$.

4._____ There are two different elements x and y in Z_7 such that $6 \cdot x = 1$ and $6 \cdot y = 1$.

5._____ For all non-zero elements a, b in Z_7, if $a = b^{-1}$ then $b = a^{-1}$.

6._____ For all a, b in Z_7, if $a \neq 0$ and $b \neq 0$ then $a \cdot b \neq 0$.

7._____ $a \cdot (b \cdot a^{-1}) = b$ for all a, b in Z_7 with $a \neq 0$.

★8._____ Find the smallest counting number n such that $a^n = 1$ for all $a \neq 0$ in Z_7.

(Note: $\underbrace{a \cdot a \cdot a \cdot \cdots \cdot a}_{n \text{ factors}} = a^n$)

★9._____ $a^7 = $_____ for all a in Z_7.

★10._____ Find an element a in Z_7 such that for every $b \neq 0$ in Z_7, $a^n = b$ for some integer n.

Note: There might be a different integer n for each b in Z_7. For example, try $a = 2$.

$$2^1 = 2 \qquad 2^4 = 2$$
$$2^2 = 4 \qquad 2^5 = 4$$
$$2^3 = 1 \qquad 2^6 = 1$$

Why is 2 not a suitable choice for a?

2.6 STATEMENTS ABOUT $(Z_7, +, \cdot)$

PROBLEM 1   Use the tables for $(Z_7, +, \cdot)$ to complete each of the following open sentences.
   a. $6 \cdot (4 + 3) = $_____
   b. $(6 \cdot 4) + (6 \cdot 3) = $_____
   c. $(6 \cdot 4) + 3 = $_____
   d. $(^-5) \cdot (^-2) = $_____
   e. $5 \cdot 2 = $_____
   f. $^-3 \cdot (2 + 4) = $_____
   g. $3 \cdot (^-2) + 3 \cdot (^-4) = $_____

Now that you've completed some open sentences in $(Z_7, +, \cdot)$, you can begin to make some generalizations.

PROBLEM 2   Try completing this generalization, if possible:

$$(^-1) \cdot a = a \cdot (^-1) = \underline{\quad\quad}, \text{ for all } a \text{ in } Z_7.$$

PROBLEM 3   For each statement that follows, decide T or F. Write *proven* if you prove that a statement is true.

a. $\underline{\quad\quad}$ $a \cdot (b + c) = a \cdot b + a \cdot c$ for all a, b, c in Z_7.

b. $\underline{\quad\quad}$ $a + (b \cdot c) = (a + b) \cdot (a + c)$ for all a, b, c in Z_7.

c. $\underline{\quad\quad}$ $(^-a) \cdot b = a \cdot (^-b) = {}^-(ab)$ for all a, b in Z_7.

d. $\underline{\quad\quad}$ $(^-a) \cdot (^-b) = a \cdot b$ for all a, b in Z_7.

e. $\underline{\quad\quad}$ $a^{-1} + b^{-1} = (a + b)^{-1}$ for all a, b in Z_7 such that a, b, and $a + b$ are nonzero.

f. $\underline{\quad\quad}$ $a^{-1} = {}^-a$ for all $a \neq 0$ in Z_7.

g. $\underline{\quad\quad}$ $a^{-1} = {}^-a$ for some a in Z_7.

Now you've considered how two operations, addition and multiplication, act simultaneously in Z_7. New possibilities occur in this case. Additive inverses are multiplied and multiplicative inverses are added. You can even rewrite an expression like $2^{-1} \cdot 3 + 2^{-1} \cdot {}^-5$ in an entirely different way: $2^{-1} \cdot (3 + {}^-5)$. New structure on Z_7 results from this interaction of multiplication and addition. The system becomes more complicated with two operations, but more interesting.

HOMEWORK EXERCISES

In Exercises 1 to 3, decide T or F. Feel free to use results of homework problems from Section 2.5.

1. $\underline{\quad\quad}$ $(a + b) \cdot c = a \cdot c + b \cdot c$ for all a, b, c in Z_7.

2. $\underline{\quad\quad}$ $(a + b) \cdot c^{-1} = a \cdot c^{-1} + b \cdot c^{-1}$ for all a, b, c in Z_7 with $c \neq 0$.

3. $\underline{\quad\quad}$ $a \cdot (b + c)^{-1} = a \cdot b^{-1} + a \cdot c^{-1}$ for all permissible elements a, b, c in Z_7. (For this statement to make sense, which elements must be nonzero?)

4. _____ **a.** Find all pairs a, b in Z_7 such that $a^2 = b^2$.
   b. Show that $(a + b)^7 = a + b$ for all a, b in Z_7.
   c. Show that $(a + b)^7 = a^7 + b^7$ for all a, b in Z_7.

2.7   FUNDAMENTAL PROPERTIES OF $(Z_7, +, \cdot)$

In the previous sections you have identified a number of true statements about the system $(Z_7, +, \cdot)$. Notice that you have a large number of these statements, and you could generate even more of them if you wished. This gives you a very large amount of information to remember about Z_7. Perhaps there is a way to organize all this information with a shorter list of statements. Fortunately, some of these statements can be regarded as *fundamental* properties of the system $(Z_7, +, \cdot)$. The remaining true statements can be derived as logical consequences of these fundamental properties.

The issue before you at present is to separate the true statements about $(Z_7, +, \cdot)$ into two categories: the fundamental properties of Z_7 and the other statements which follow from these fundamental properties.

PROBLEM 1   To facilitate your work, list the statements known to be true about (a) addition in Z_7, (b) multiplication in Z_7, and (c) both operations in Z_7. (Do not include statements examined for homework.)

PROBLEM 2   Look for logical relationships among these statements. Select as fundamental those properties which cannot, in your opinion, be derived from the other statements on your list. Check your results with your teacher.

2.8   THE FIELD PROPERTIES

The system $(Z_7, +, \cdot)$ consists of a set and two operations that satisfy certain basic properties. In the next section you will encounter several other systems with two operations denoted $+$ and $\cdot$. Some of these systems will have the same properties as $(Z_7, +, \cdot)$; others will not. $(Z_7, +, \cdot)$ and other systems having these same properties are called *fields*.

Consider, for example, the set of numbers $\{\ldots, ^-2, ^-1, 0, 1, 2, \ldots\}$ known to you from elementary school days. When you combine the members of this set under the operations of addition, subtraction, multiplication, and division, you obtain a bewildering variety of results, so enormous and seemingly chaotic that you seek ways to discover the underlying order and design. Definitions are tools used for this purpose, and the definition of a field will help you to recognize the structure of this system as well as other mathematical systems.

In your exploration of $(Z_7, +, \cdot)$ properties, you have identified one list of basic properties from which others can be proven. But there are several possible choices for such a basic list. Mathematicians have generally agreed on a list of eleven basic field properties. While the list given as Figure 2.1 might not be the same as yours, it should be possible to prove all of your properties from this "official" list.

PROBLEM 1   There are other properties of a field. These other properties can be justified in terms of the basic properties chosen as the definition.

Explain why the property

$$(a + b) \cdot c = (a \cdot c) + (b \cdot c) \text{ for all } a, b, c \text{ in } F$$

holds in a field $(F, +, \cdot)$.

Does the property

$$(s \cdot t) + (u \cdot t) = (s + u) \cdot t \text{ for all } s, t, u \text{ in } F$$

hold in a field $(F, +, \cdot)$? Justify your answer.

PROBLEM 2   Explain why the property $(x + z) + y = x + (y + z)$ holds for all x, y, z in F.

The same property holds for multiplication $(x \cdot z) \cdot y = x \cdot (y \cdot z)$ for all x, y, z in F. Justify this one.

★PROBLEM 3   The field properties can be stated more than one way. For example, the associative and commutative properties of addition can be replaced by the following property.

For all x, y, z in F, $(x + z) + y = x + (y + z)$

Using this new property and also any of the other field properties that are needed, prove the associative and commutative properties.

32

DEFINITION: Let F be a set with two operations called addition ($+$) and multiplication ($\cdot$). The system $(F, +, \cdot)$ is a field if and only if the following eleven properties are satisfied. (Complete the statements of the multiplication properties.)

Addition

1. F is CLOSED under addition:
   For all a, b in F, $a + b$ is in F.

2. Addition is ASSOCIATIVE in F:
   For all a, b, c in F, $a + (b + c) = (a + b) + c$.

3. F has an IDENTITY element for addition:
   There is an element denoted 0 in F such that $a + 0 = 0 + a = a$ for all a in F.

4. Each element in F has an ADDITIVE INVERSE in F:
   For each a in F, there is an element denoted ^-a in F such that $a + {}^-a = {}^-a + a = 0$.

5. Addition is COMMUTATIVE in F:
   For all a, b in F, $a + b = b + a$.

Multiplication

6. F is CLOSED under multiplication:
   For all a, b in F, _____ .

7. Multiplication is ASSOCIATIVE in F: For all a, b, c in F, _____ .

8. F has an IDENTITY element for multiplication:
   There is an element denoted 1 in F such that $a \cdot 1 =$ ___ $=$ ___ for all a in F.

9. Each nonzero element in F has a MULTIPLICATIVE INVERSE in F:
   For each $a \neq 0$ in F, there is an element denoted a^{-1} in F such that $a \cdot a^{-1} = a^{-1} \cdot a =$ _____ .

10. Multiplication is COMMUTATIVE in F:
    For all a, b in F, _____ .

11. Multiplication is DISTRIBUTIVE over addition in F:
    For all a, b, c in F,
    $a \cdot (b + c) = (a \cdot b) + (a \cdot c)$.

FIGURE 2.1

Consider the nature of your work so far. You've inductively explored a *specific* mathematical system $(Z_7, +, \cdot)$ and developed a list of basic properties of this system. Then you've generalized these basic properties to a system $(F, +, \cdot)$ called a field, where F is *any* set satisfying these basic properties under addition and multiplication. You have just used deductive processes to justify

33

additional properties in a field. Soon you'll be using deductive methods again by testing whether or not different sets form fields under addition and multiplication.

HOMEWORK EXERCISES

1. In your previous work with Z_7, have you shown that $(Z_7, +, \cdot)$ is a field, i.e. that it satisfies all the field properties?

★2. The associative and commutative properties for multiplication can be deleted from the field properties, in the same way as those for addition were deleted in Problem 3, and replaced by the following property:

For all x, y, z in F, $(x \cdot z) \cdot y = x \cdot (y \cdot z)$

Using this new property and also any of the other field properties that are needed, prove the associative and commutative properties of multiplication.

2.9   PROPERTIES OF WHOLE NUMBERS, INTEGERS, AND RATIONALS

The eleven properties that define a field are basic because many other properties of + and $\cdot$ are logical consequences of this fundamental list. Refer to Problem 2 in Section 2.8. Perhaps the properties are also practical tools for simplifying arithmetic calculation with whole numbers. For example, in the problem $(115 + 273) + 85$, the parentheses dictate beginning by calculating

$$115 + 273 = 388$$

and then $388 + 85 = 473$. But if whole number addition is commutative and associative, as in the field $(Z_7, +, \cdot)$, the calculation could be done by noting that

$$
\begin{aligned}
(115 + 273) + 85 &= (273 + 115) + 85 \\
&= 273 + (115 + 85) \\
&= 273 + 200 \\
&= 473
\end{aligned}
$$

Of course, the rearrangement could be done mentally!

Even more work perhaps could be saved in a multiplication problem such as

$$(2005) \cdot 25 = (2000 \cdot 25) + (5 \cdot 25)$$
$$= 50000 + 125$$
$$= 50125$$

All of that calculation could really be done mentally. But unless $(W, +, \cdot)$ has a distributive property of multiplication over addition, the answer arrived at by the shortcut might not be correct.

We will look next at some familiar number systems: whole numbers, integers, and rational numbers. We will briefly recall the arithmetic in these systems and then determine which of the field properties hold in each of them.

The set of *whole numbers* is named by *W*. $W = \{0, 1, 2, 3, 4, \ldots\}$

PROBLEM 1   What do you perceive to be the essential difference between the *sets W* and Z_7?

The set of *integers* is named by *Z*. $Z = \{\ldots, {}^-5, {}^-4, {}^-3, {}^-2, {}^-1, 0, 1, 2, 3, 4, 5, \ldots\}$

PROBLEM 2   Describe the relationship of set *W* to set *Z*.

PROBLEM 3   To check your recall of integer arithmetic, perform each of the following calculations in $(Z, +, \cdot)$.
   a. $1202 + 31 = $ _____
   b. $1202 + {}^-31 = $ _____
   c. ${}^-1202 + {}^-31 = $ _____
   d. ${}^-1202 + 31 = $ _____
   e. $37 \cdot 24 = $ _____
   f. ${}^-37 \cdot 24 = $ _____
   g. $37 \cdot {}^-24 = $ _____
   h. ${}^-37 \cdot {}^-24 = $ _____

The set of *rational numbers*, named by *Q*, includes all quotients of integers with denominator *never* zero. Rational numbers are named by common fractions like $\frac{2}{5}$, $\frac{5}{8}$, and $\frac{4}{1}$ or by decimals like 0.4, ${}^-0.625$, ${}^-1.5$, and $0.666\ldots$.

The set of *rationals* greater than zero is denoted by Q^+.

PROBLEM 4   Describe the relationship of:
   a. set W to set Q
   b. set W to set Z
   c. set Z to set Q^+
   d. set Z to set Q
   e. set Q^+ to set Q

PROBLEM 5   To check your recall of rational number arithmetic, perform each of the following calculations in $(Q, +, \cdot)$.

   a. $\dfrac{2}{7} + \dfrac{3}{7} =$ _____

   b. $\dfrac{1}{2} + \dfrac{^-1}{3} =$ _____

   c. $\dfrac{^-3}{5} + \dfrac{5}{6} =$ _____

   d. $\dfrac{^-3}{7} + \dfrac{^-2}{7} =$ _____

   e. $\dfrac{1}{2} \cdot \dfrac{1}{3} =$ _____

   f. $\dfrac{1}{2} \cdot \dfrac{^-1}{3} =$ _____

   g. $\dfrac{1}{2} \cdot \dfrac{5}{6} =$ _____

   h. $\dfrac{^-8}{7} \cdot \dfrac{^-2}{7} =$ _____

   i. $(2.7) \cdot (1.6) =$ _____

   j. $(1.32) \cdot (2.01) =$ _____

   Having reviewed the arithmetic operations of $+$ and $\cdot$ in W, Z, and Q, it is of interest whether $(W, +, \cdot)$, $(Z, +, \cdot)$, and $(Q, +, \cdot)$ are fields.

   Recall that in $(Z_7, +, \cdot)$ you can (in principle) prove that the field properties are true by examining all possible instances of each property. For example, you can prove that Z_7 is closed under addition by computing all possible sums of two elements in Z_7 and noting that each result is an element of Z_7. This is another example of proof by exhaustion.

PROBLEM 6   Can the method of proof by exhaustion be used to show the truth of a field property in W, Z, or Q? Why or why not?

PROBLEM 7   Now use Table 2.7 to record your results in determining whether W, Z, and Q are fields. For each property and each system, determine in your group whether the property is true or false. If a property holds in a given system, write T in the appropriate spot in the table. However, if a property does not hold in the given system, write F and give one numerical *counterexample* (that is, one instance for which the property fails) in Table 2.7.

TABLE 2.7

Field Property	W	Z	Q^+	Q
Closure under +				
Associativity for +				
Additive Identity				
Additive Inverse				
Commutativity for +				
Closure under ·				
Associativity for ·				
Multiplicative Identity				
Multiplicative Inverse for all Nonzero Elements				
Commutativity for ·				
Distributivity of · over +				

a. For each of the systems W, Z, Q^+, and Q identify those field properties (if any) which fail.
  In W, _____
  In Z, _____
  In Q^+, _____
  In Q, _____
b. Which of these systems are fields?
c. For which systems have you *proved* that your conclusion is correct?

PROBLEM 8   a. To determine if $(S, +, \cdot)$ is a field, investigate the properties of $(S, +, \cdot)$ where S is the set of rational numbers between $^-1$ and 1 inclusive:

$$S = \left\{ {}^-1, \ldots, \frac{{}^-1}{3}, \ldots, 0, \ldots, \frac{1}{2}, \ldots, 1 \right\}$$

b. Specify those properties, if any, which fail.

HOMEWORK EXERCISES

Investigate the properties of each of the following systems to determine if it is a field. Specify those properties, if any, which fail.

1. $(N, +, \cdot)$ where N is the set of *natural numbers*:

$$N = \{1, 2, 3, \ldots\}$$

Note: The set of natural numbers is the same as the set of counting numbers.

2. $(E, +, \cdot)$ where E is the set of even whole numbers:

$$E = \{0, 2, 4, \ldots\}$$

3. $(P, +, \cdot)$ where P is the set of nonnegative powers of 2:

$$P = \{1, 2, 4, 8, \ldots\}$$

Before investigating the properties of $(P, +, \cdot)$, find all elements of P less than 40.

2.10 SOLUTION OF EQUATIONS IN VARIOUS SYSTEMS

You have previously identified a number of similarities and differences among the properties of the systems N, W, Z, Q, Z_7. In this section you will examine linear equations in these systems. You will see how the properties of various systems affect the possibility of solving equations in those systems.

PROBLEM 1   In $(Z_7, +, \cdot)$: Does the equation $x + 5 = 3$ have a solution for x?

$$x + 2 = 5?$$
$$x + 1 = 1?$$
$$x + 6 = 3?$$

PROBLEM 2   Answer Problem 1 for sets N, W, Z, and Q. List each set in which the given equation has a solution, and find the solution when it exists. **Note:** Set N is described in Section 2.9, Homework Exercise 1.

PROBLEM 3   Consider many *linear equations* of the form $x + b = c$, where b and c both belong to the same set. (Keep in mind that b may equal c, as in one of the equations in Problem 1.) Decide which sets among Z_7, *N, W, Z, Q* have the following property:

For every linear equation $x + b = c$ with b and c *in the given set* there is a solution for x *in the given set.*

PROBLEM 4   Record your results from Problem 3 in Table 2.8, in this manner: For each set which lacks the stated property, give a counterexample, i.e., an equation $x + b = c$ with b and c *in the given set* but with no solution for x *in the given set.* As an example, suppose b and c are in N with $b = c = 5$, then does $x + b = c$ have a solution in N? For each set which has the stated property, provide an example of an equation and its solution in the given set. Make your example different from the equations in Problem 1.

TABLE 2.8

	Examples: If the Property Always Holds	*Counterexamples: If the Property Fails*
Z_7		
N		
W		
Z		
Q		

In which sets among Z_7, *N, W, Z, Q* is it possible to solve *every* linear equation $x + b = c$, where b and c are in the given set?

PROBLEM 5   In each of the appropriate sets, search for what you think is the solution of the equation $x + b = c$, with b and c in the given set $x =$ _____. (If your answer involves the operation of subtraction, change it to an answer involving an additive inverse.)

Now substitute this solution in the equation for x to check that this really is a solution for the equation. This testing to see if a solution exists is called an *existence proof.*

What properties are used in this existence proof?

PROBLEM 6   Now consider linear equations of the form $a \cdot x = c$, where $a \neq 0$, and a and c belong to the same set: Z_7, N, W, Z, or Q. Decide which of these systems have the following property:

For every linear equation $a \cdot x = c$ with a and c in the given set and $a \neq 0$, there is a solution for x in the given set.

Record your results in Table 2.9, using directions as in Problem 4.

TABLE 2.9

	Examples	Counterexamples
Z_7		
N		
W		
Z		
Q		

In which sets is it possible to solve every linear equation $a \cdot x = c$ with $a \neq 0$ and both a and c in the given set? _____

PROBLEM 7   In each of the appropriate sets, search for what you think is a solution of the equation $a \cdot x = c$ with $a \neq 0$. $x =$ _____. (If your answer involves the operation of division, change it to an answer involving a multiplicative inverse.)

Now substitute this solution in the equation for x to check that this really is a solution for the equation.

What properties are used in this existence proof?

PROBLEM 8   If a, b, and c are elements of a field, what do you think is a solution of the equation $a \cdot x + b = c$? $x =$ _____.

Substitute this solution in the equation for x.

What properties are used in checking the existence of this solution?

Again you have used an inductive process by looking for solutions of many specific linear equations in different systems and then considering a general equation. You've decided upon the shared properties of those systems which always had solutions to linear equations. Now, deductively, you can check that a solution must be unique if these properties are satisfied by a given system.

PROBLEM 9   In Problem 8, you found that there is one solution for the equation $ax + b = c$. In a field, can there be other solutions for x? Justify your answer.

★PROBLEM 10   Formalize your justification in Problem 9 and in so doing, prove that the equation $ax + b = c$, $a \neq 0$ has exactly one solution for x.

PROBLEM 11   Find general solutions for equations in Q of the four types given. In each case determine the conditions affecting the existence of solutions and the possibilities for solutions for each situation.

a. $ax = b$

b. $ax + b = cx + d$

c. $\dfrac{ax + b}{c} = d$

d. $\dfrac{ax + b}{c} = \dfrac{dx + e}{f}$

HOMEWORK EXERCISES

1. Consider equations of the form $a \cdot x + b = c$. Decide which of the sets Z_7, N, W, Z, Q have the following property.

   For every linear equation $ax + b = c$ with a, b, c in the given set and $a \neq 0$, there is a solution for x in the given set. Make a chart as before. Then answer the following question from your chart: In which sets is it possible to solve every linear equation $ax + b = c$ with a, b, c in the given set and $a \neq 0$? _____ Note that the multiplication sign was dropped in representing the linear equation $ax + b = c$. Multiplication is often indicated in this simple way.

2. For each equation given below, determine whether the equation has a solution in the given system. If so, find the solution. **Note** that although not *every* linear equation has a solution in a particular set, *some* linear equations may have a solution. For example, $5x = 9$ has no solution in W, but $5x = 10$ does.

   In Z_7:   a. $5x + 1 = x + 4$

$\quad\quad\quad\quad$ **b.** $^-2 \cdot x + ^-6 = ^-4 \cdot x + ^-5$

$\quad\quad\quad\quad$ **c.** $x + ^-4 = 3x + ^-2$

In W: $\quad$ **a.** $3x + 151 = 172$

$\quad\quad\quad\quad$ **b.** $2x + 46 = 38$

In Z: $\quad$ **a.** $5x + 98 = 53$

$\quad\quad\quad\quad$ **b.** $7x + 73 = 102$

In Q: $\quad$ **a.** $11x + 11 = 6x + 8$

$\quad\quad\quad\quad$ **b.** $4x + 5 = \dfrac{5}{2}x + 7$

$\quad\quad\quad\quad$ **c.** $\dfrac{1}{2}x + \dfrac{7}{6} = \dfrac{2}{3}x + \dfrac{3}{5}$

3. Determine the conditions for existence of solutions to the equation $ax + b = c$ in Q if we do not stipulate that $a \neq 0$. Would there ever be a solution if $a = 0$? If so, how many?

2.11 FLOWCHARTS FOR SOLUTIONS OF EQUATIONS

In the preceding section you developed a general procedure or *algorithm* for solving equations of the form $ax = b$. The virtue of this general solution is that the task of solving any specific equation of this type can now be delegated to a person or machine capable only of directed arithmetic computation.

After solving the problem and getting the solution $x = a^{-1} \cdot b$, you are ready to tell your assistant (whether human or mechanical) exactly which steps to perform in order to calculate the solution for a given equation.

Notice that there are two different phases to the entire procedure. First you analyze and solve the problem generally. Second, you give instructions to the assistant for finding the solutions to specific problems. Your assistant *does not* actually *solve* the problem, but focuses on the solution that you have found. This idea of telling your assistant what steps to perform is done easily with a flowchart.

PROBLEM 1 $\quad$ Construct a flowchart to find the solution of the equation $ax = b$, allowing for each condition that affects both the existence of solutions

42

and possibilities for solutions in each situation. Use your analysis from Section 2.10, Homework Exercise 3.

HOMEWORK EXERCISES

Construct flowcharts for each of the following tasks.

1. Solve equations of the form $ax + b = cx + d$.

2. Solve equations of the form $\dfrac{ax + b}{c} = \dfrac{dx + e}{f}$

3. Add fractions and express the answer as a fraction. Assume that the computer can only add, subtract, and multiply with integers.

4. Multiply fractions and express the answer as a fraction.

5. Add any two numbers in $(Z_7, +)$.

6. Multiply any two numbers in $(Z_7, \cdot)$.

2.12 PROGRAMMING IN THE BASIC LANGUAGE

Now that you have a flowchart for solving equations of the form $ax = b$, the next step is to actually write a computer program, using the BASIC language, to solve such equations.

The BASIC language has a very limited set of instructions and is thus relatively simple to learn and use. Each program statement performs one instruction and must begin with a counting number between 1 and 99999 (inclusive). The computer executes statements in their *numerical* order unless otherwise directed in the program.

Figure 2.2 shows a very simple flowchart for solving the equation $ax = b$. Notice that this flowchart does *not* allow for the possibility that $a = 0$. It is being used because it provides an easy introduction to the use of the BASIC language.

According to the flowchart in Figure 2.2, the first step for the computer is to receive values for a and b. (Since the teletype

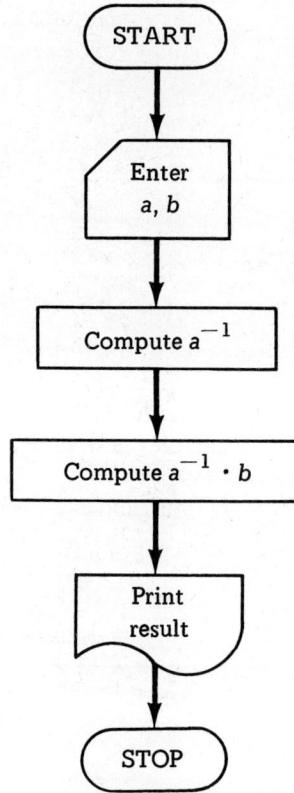

FIGURE 2.2

cannot type lower case letters, A and B must be used.) This can be done with an INPUT statement as follows:

```
10 INPUT A, B
```

The INPUT statement instructs the computer to set up storage locations called A and B. When the program is executed (run), the computer will type a question mark, ?, and expect two numbers separated by a comma to be typed on the teletype by the program user. The first number will be stored in the location called A, and the second number will be stored in the location called B.

The flowchart indicates that the next step of the program is to compute A^{-1}. Computations are done with LET statements.

The teletype has no provision for either subscripts or superscripts, so to indicate an exponent, a double asterisk, ∗∗, is used. For example, 3^2 is written 3∗∗2 and 5^3 is written 5∗∗3. (On some teletypes and computers, an up-arrow, ↑, serves this purpose.)

20 LET Z=A∗∗-1

Note: The choice of Z is arbitrary. Any letter could have been used *except* A and B which have already been used in the program.

Since $A^{-1} = 1/A$, the following statement could also have been used:

20 LET Z=1/A

The slash, /, is used to represent division on the computer.

The next step is to multiply A^{-1} and B, and store the result in a location called X. A single asterisk, ∗, is used to denote multiplication on the computer. All operations must be specifically indicated in the program statements. For example, if you wish to multiply D and E, DE is *not* valid; D∗E must be used. (For addition and subtraction, the usual symbols, + and −, are used.)

30 LET X=Z∗B

The computer now has the numerical solution to this equation stored in the location called X. The last thing to do is print out this value. To print out something from the computer, a PRINT statement is used.

40 PRINT X

Every BASIC program must terminate with an END statement to signify the end of the program.

50 END

The flowchart and complete program are shown in Figure 2.3.

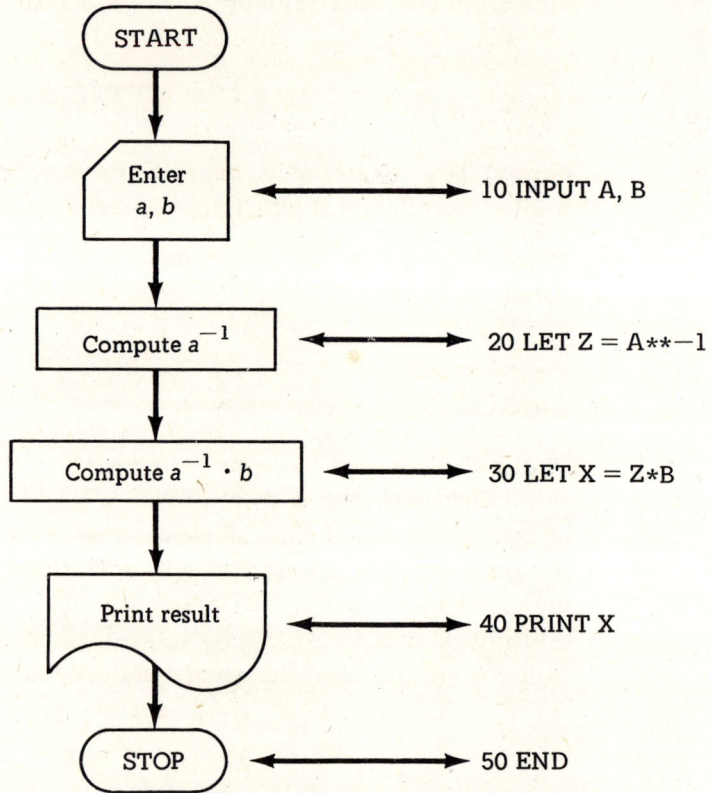

FIGURE 2.3

Note: The use of blanks in program statements is optional and does not affect the program. Thus, the statements 10IN PUTA, B; 10INPUTA,B; and 10 INPUTA,B are all equivalent. However, to make reading the statements easier, the use of blanks is recommended.

This program does not allow for special handling of the case where A = 0. The program will be modified later to incorporate the contingencies flowcharted in the previous section, but first another option will be considered.

The program in Figure 2.3 is, admittedly, a lot of work just to solve one very simple equation. This program does not begin to use the power of the computer and was used only to illustrate some of the statements in the BASIC programming language. The program can be made more useful if it can be altered to calculate solutions for many equations. Fortunately, this is done very easily. Figure 2.4 shows a modified flowchart and program to accomplish this task.

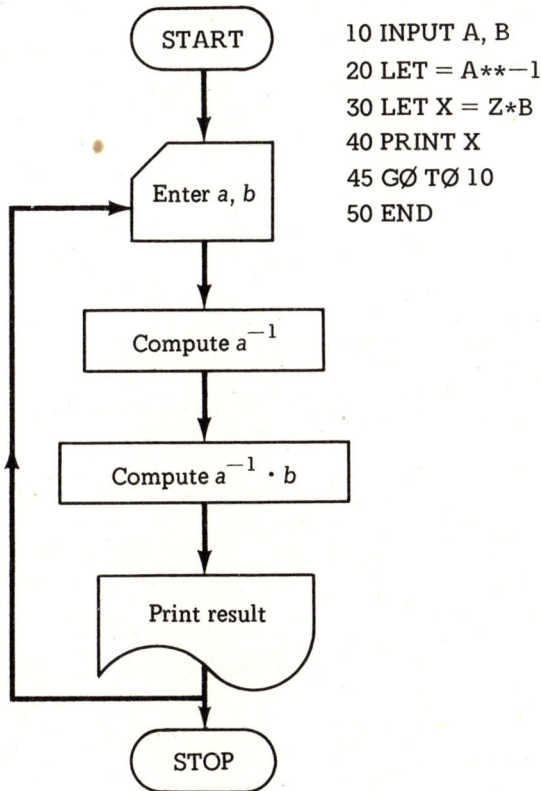

```
10 INPUT A, B
20 LET = A**−1
30 LET X = Z*B
40 PRINT X
45 GØ TØ 10
50 END
```

FIGURE 2.4

The GØ TØ statement is used to change the numerical order of execution of the program statements.

Note: Typing in all of the program statements in your program is equivalent to acquainting your assistant with the procedure you want executed. To actually execute (run) the program, the command RUN must be typed.

Executing the program in Figure 2.4 to solve the three equations $3x = 2$, $5x = 10$, $^-7x = 6$ would yield the following output (underlining is used here to indicate what the computer prints although the computer will not underline). **Note:** The computer does not print fractions.

RUN

<u>?</u> 3, 2

<u>.666667</u>

<u>?</u> 5, 10

<u>2.0</u>

<u>?</u> -7, 6

<u>-.857143</u>

<u>?</u> STØP

<u>PRØGRAM STØPPED</u>

<u>TIME .023</u>

HOMEWORK EXERCISES

1. Correct each of the following BASIC expressions.
   a. $A \div L$
   b. $2(L + W)$
   c. $\frac{1}{2}BH$
   d. $2\frac{1}{2} * 3\frac{1}{4}$
   e. $(7 + 8)/2$

2. Find *all* the mistakes in each of the following BASIC programs.

a.   10 PRNT 3.1416*2.71828**2
     20 END
     30 RUN

b.   10 PRINT 6(5.280+333)
      9 END
       RUN

c.   10 LET L=17
     20 LET A=L*W
     30 PRINT A
     END
     RUN

d.   10 LET L=M
     20 LET 12=W
     30 LET A=L*W
     40 PRINT A
     29 END
     RUN

3. Find the output of the following BASIC programs.

a.   10 LET X=4
     20 LET Y=(X**2)*X
     30 LET Z=Y-X
     40 PRINT Z
     50 END

b.   100 LET A=2
     150 PRINT A
     200 LET A=A+A
     250 PRINT A
     300 LET A=A*A
     350 PRINT A
     400 LET A=A**2
     450 PRINT A
     500 END

c.   10 INPUT A,B,C,D
     20 LET E=A*D-B*C
     30 PRINT E
     40 LET F= -(A/E)

```
50 PRINT F
60 END
RUN
? 1, 5, 4, 25
```

d.
```
20 INPUT X,Y,Z
30 LET S=X+Y
40 LET T=Z+S
50 LET S=T/(4+X-(1+X))
60 PRINT S
70 END
RUN
? 1, 2, 4
```

4. Write a program in the BASIC language that will solve an equation of the form $ax + b = cx + d$.

5. Use your program from the previous problem to solve the following three equations on the computer.
   a. $3x + 7 = 2x + 17$
   b. $^-11x + (^-4) = 9x + 76$
   c. $.85x + 1.02 = 1.15x + 17.52$

2.13 REFINING PROGRAMS

The program already written to solve the equation $ax = b$ does not allow for the case in which $a = 0$. To allow for this possibility, consider the flowchart in Figure 2.5. Other flowcharts are possible, but they should all contain essentially the same steps. You may want to compare this flowchart with the one you wrote in Section 2.11 to see that they both convey the same idea.

Again, the program is written directly from the flowchart. The first statement is the same as in the previous program.

```
10 INPUT A,B
```

The next statement according to the flowchart must decide

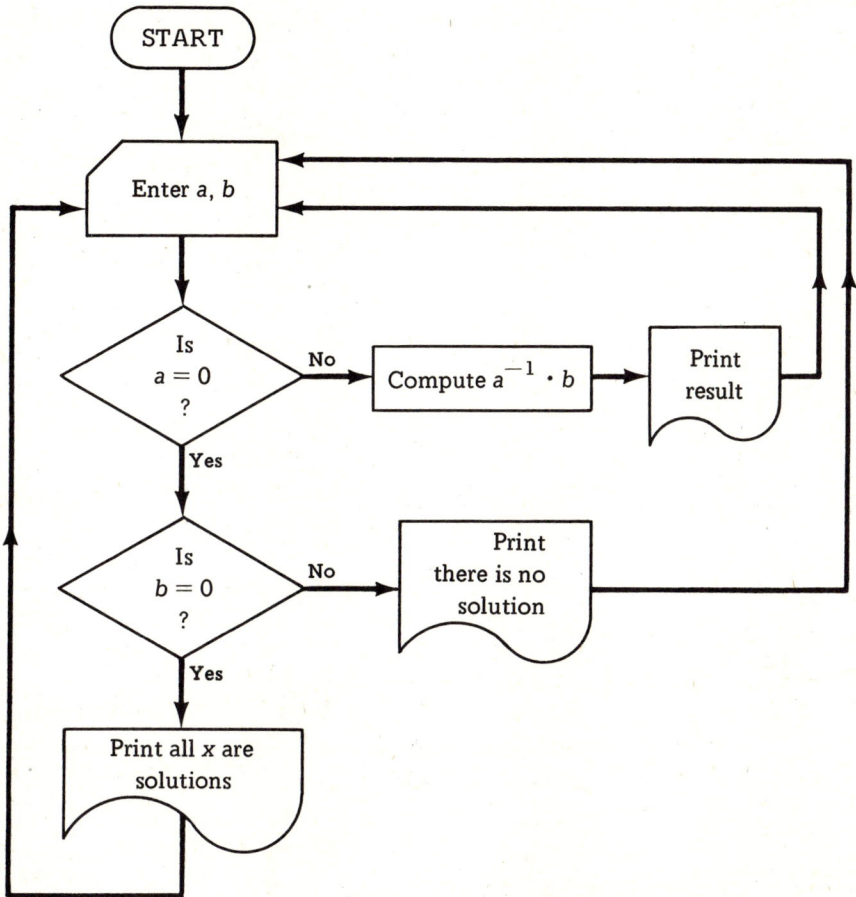

FIGURE 2.5

whether or not $A = 0$. Decision boxes in a flowchart are represented by IF ... THEN ... statements in the corresponding programs. The statement used in this program is:

```
20 IF A=0 THEN 60
```

The function of the IF ... THEN ... statement is very logical. If
$A = 0$ then the computer executes statement 60 next. If $A \neq 0$ the
computer acts as if Statement 20 is not even in the program and
executes the next statement (numerically) after Statement 20.

Table 2.10 indicates other relational symbols that can be used in IF . . . THEN . . . statements.

TABLE 2.10

Algebraic Relation	Teletype Symbol
$=$	$=$
$<$	$<$
$>$	$>$
$\neq$	$><$ or $<>$
$\leqslant$	$<=$ or $=<$
$\geqslant$	$>=$ or $=>$

The next statement in the program represents the path taken when $A \neq 0$.

```
30 LET X=(A**-1)*B
```

The LET statement has already been introduced; however, this statement illustrates the fact that a LET statement can do more than one calculation. This can often substantially reduce the number of statements in a program. In this case, two calculations, computing A^{-1} and multiplying that result by B, are combined in one statement.

To print out the result, the PRINT statement is used.

```
40 PRINT 'THE SØLUTIØN IS'; X
```

To make the output a little easier to read, another aspect of the PRINT statement is illustrated. Any string of characters contained between single quotes (apostrophes) in a PRINT statement is printed exactly as entered. The semicolon is used to signal the computer to print the *value* of X on the same line as the phrase enclosed in quotes.

The rest of the program does not involve any new ideas and can be written by following the flowchart. The complete program containing eleven lines follows:

```
 10  INPUT A,B
 20  IF A=0 THEN 60
 30  LET X=(A**-1) *B
 40  PRINT 'THE SØLUTIØN IS'; X
 50  GØ TØ 10
 60  IF B=0 THEN 90
 70  PRINT 'THERE IS NØ SØLUTIØN'
 80  GØ TØ 10
 90  PRINT 'ALL X ARE SØLUTIØNS'
100  GØ TØ 10
110  END
```

Using the preceding program to solve the equations $0x = 5$, $0x = 0$, $3x = 2$, and $2x = 0$ would yield the following output.

```
RUN

? 0,5

THERE IS NØ SØLUTIØN

? 0,0

ALL X ARE SØLUTIØNS

? 3,2

THE SØLUTIØN IS .666667

? 2,0

THE SØLUTIØN IS 0

? STØP

PRØGRAM STØPPED

TIME .040
```

One problem with this program is that the person using it must know to input two values when the computer prints a question mark. To remedy this shortcoming and thus make the program useful to people unfamiliar with the program, the PRINT statement can be used to add some explanation of the program's purpose and use.

For example, the following statements could be added:

```
4 PRINT 'THIS PRØGRAM SØLVES EQUATIØNS ØF THE'
5 PRINT 'FØRM AX=B'
6 PRINT
7 PRINT 'ENTER THE VALUES ØF A AND B IN ØRDER'
8 PRINT 'AND SEPARATED BY A CØMMA'
```

Statement 6 will not print anything and only serves to skip a line in the output. By changing Statements 50, 80, and 100 to GØ TØ 6, instructions will be printed out before each question mark.

The entire program, together with sample output is given in Figure 2.6.

Note: By numbering the original program with multiples of ten, it was possible to revise the program by simply inserting additional program statements anywhere in the program merely by using an appropriate statement number.

```
  4 PRINT 'THIS PRØGRAM SØLVES EQUATIØNS ØF THE'
  5 PRINT 'FØRM AX=B'
  6 PRINT
  7 PRINT 'ENTER THE VALUES ØF A AND B IN ØRDER'
  8 PRINT 'AND SEPARATED BY A CØMMA'
 10 INPUT A, B
 20 IF A=0 THEN 60
 30 LET X=(A**-1)*B
 40 PRINT 'THE SØLUTIØN is';X
 50 GØ TØ 6
 60 IF B=0 THEN 90
 70 PRINT 'THERE IS NØ SØLUTIØN'
 80 GØ TØ 6
 90 PRINT 'ALL X ARE SØLUTIØNS'
100 GØ TØ 6
110 END
```

```
RUN
THIS PRØGRAM SØLVES EQUATIØNS ØF THE
FØRM AX = B

ENTER THE VALUES ØF A AND B IN ØRDER
AND SEPARATED BY A CØMMA.
? 0, 5
THERE IS NØ SØLUTIØN

ENTER THE VALUES ØF A AND B IN ØRDER
AND SEPARATED BY A CØMMA.
? 0, 0
ALL X ARE SØLUTIØNS

ENTER THE VALUES ØF A AND B IN ØRDER
AND SEPARATED BY A CØMMA.
? STØP
PRØGRAM STØPPED

TIME .041
```

FIGURE 2.6

Table 2.11 summarizes the BASIC language programming statements introduced up to this point.

TABLE 2.11   Summary of BASIC Statements

Statement	Function	Example
INPUT	Reads data from teletype	10 INPUT A, B, C
LET	Computes and assigns value	35 LET X = (A + B)*C
PRINT	Types out values and messages	40 PRINT 'SOLUTION IS'; X
IF . . . THEN . . .	Conditional transfer	60 IF X=<40 THEN 100
GØ TØ	Alters program flow	51 GØ TØ 10
END	Final statement in program	7000 END

HOMEWORK EXERCISES

1. Write programs to solve each of the following sets of equations (three separate programs). **Hint**: Refer to the flowcharts you made for solving each type of equation.

   a. $3x = 7$
   $$^-43x = ^-37$$
   $$.7x = ^-.02$$

   b. $\dfrac{9x + 2}{12} = 36$

   $$\dfrac{5943x + 1000}{^-104} = 119$$

   $$\dfrac{^-11x + (^-6)}{7} = ^-23$$

   c. $\dfrac{^-2x + 9}{3} = \dfrac{7x + 1}{4}$

   $$\dfrac{x - 2}{2} = \dfrac{x - 2}{3}$$

   $$\dfrac{19x + (^-6)}{7} = 0$$

2. How can the program developed in Exercise 1c be used for the programs in Exercises 1a and 1b?

3. What output would you get for the program in Figure 2.6 if you entered the values 0, 3 for *A, B* respectively and omitted Statement 80 from the program?

4. Refer to your flowchart for finding sums in mod 7 and write a program to compute such sums. Use appropriate INPUT and PRINT statements to make the program usable by other people.

5. Proceed as for Exercise 4, but have the program compute mod 7 products.

6. Combine Exercises 4 and 5 into a single program which prints out the sum and product of any two numbers in $(Z_7, +, \cdot)$.

7. Write a flowchart and program for finding additive and multiplicative inverses in $(Z_7, +, \cdot)$. Include appropriate messages for situations where inverses do not exist.

8. Modify your programs in Exercises 5 and 6 to work for *any* mod.

2.14 SUBTRACTION IN VARIOUS SYSTEMS

When you were solving linear equations in Z_7 you might have been tempted to use subtraction and division. However, there is a problem. You have not defined either subtraction or division in Z_7. Once you have defined subtraction and division in Z_7 you will be able to examine the systems $(Z_7, -)$ and $(Z_7, \div)$ and see what properties they have.

PROBLEM 1   You can start this investigation by defining subtraction in Z_7. There are a number of ways to do this. Think of what you want subtraction to do; then state a definition or procedure for subtracting any two elements of Z_7.

You might start by trying the following examples in Z_7.

$6 - 2 = $ _____

$5 - 0 = $ _____

$4 - 4 = $ _____

$0 - 2 = $ _____

$4 - 6 = $ _____

$1 - 5 = $ _____

Definition of Subtraction in Z_7: _____

When you think you have a definition, write it and ask your instructor to check it.

PROBLEM 2   Does your definition of subtraction work for all numbers in Z_7?

PROBLEM 3   Does your definition of subtraction work in Q? in Z?

There are many possibilities for the definition of subtraction in Z_7. You may have come up with this definition:

$$a - b = a + {}^-b \text{ for all } a, b \text{ in } Z_7$$

PROBLEM 4   Try using this definition on some subtraction problems in Z_7.

$3 - 5 = 3 + (\quad) = $ _____

$1 - 6 = $ _____

$5 - 2 = $ _____

$4 - 4 = $ _____

$0 - 3 = $ _____

PROBLEM 5   Could this definition of subtraction be used for all a, b in Z? for all a, b in Q?

You may have chosen another possibility for the definition of subtraction in Z_7: for all a, b in Z_7 $a - b = x$ if and only if $x + b = a$

Note: This means, if $a - b = x$ then $x + b = a$ *and* if $x + b = a$ then $a - b = x$.

PROBLEM 6   Try using this definition on some subtraction problems in Z_7.
$3 - 4 = x, x + 4 = 3, x =$ _____
$2 - 5 = x,$ _____
$6 - 3 = x,$ _____
$0 - 6 = x,$ _____
$1 - 1 = x,$ _____

PROBLEM 7   Can you apply this definition of subtraction in sets Z and Q? Now look at the whole numbers.

PROBLEM 8   Can either of the two definitions of subtraction given above be applied in W?

If a and b are elements of W, what restrictions must be placed on a and b so that $(a - b)$ is an element of W?

You have developed and examined two alternative definitions for subtraction in Z_7. There is generally more than one correct approach to a situation in mathematics. Such is the richness of the subject.

HOMEWORK EXERCISES

You now have two definitions of subtraction. Another possible definition in Z_7 is to think of the seven elements on a clock and run the clock forward for addition and backward for subtraction.

For example, $4 + 5 = 2$ in $(Z_7, +)$.

1.   Use the clock to illustrate

$$2 - 5 = 4 \text{ in } (Z_7, -).$$

First move

Second move

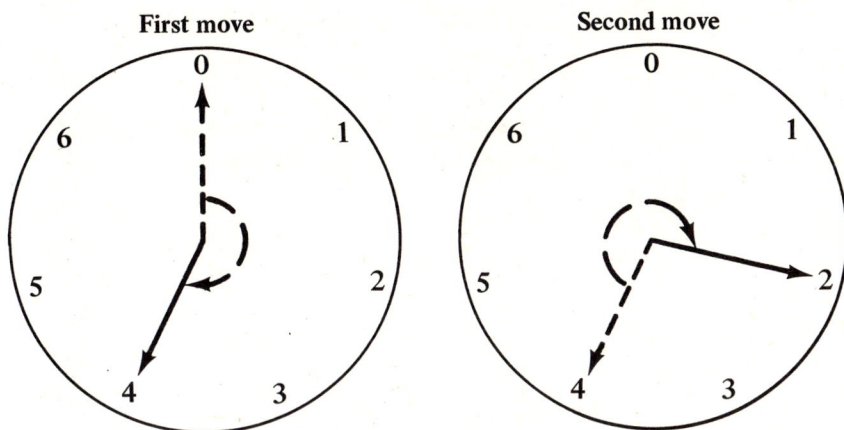

FIGURE 2.7

2. Use the clock scheme to complete the following subtraction table. Table 2.12 is to be read, as usual, with the first number read from the left side and the second number read from the top. For example, the circled 3 in the first row is the answer to $0 - 4$ in Z_7.

TABLE 2.12

−	0	1	2	3	4	5	6
0	0	6			③		
1						3	
2	2				5		
3			1	0	6		
4	4					6	
5	5			2			
6		5			2	1	

3. Use the $(Z_7, -)$ table to complete the following open sentences.
   a. $5 - 2 =$
   b. $6 - 4 =$
   c. $3 - 5 =$
   d. $2 - 6 =$

4. Now do the same four examples in Exercise 3 using the definition $a - b = a + {}^-b$ for all a, b in Z_7.

Example

$5 - 2 = 5 + {}^-2 = 5 + 5 =$

5. Do the same four examples in Exercise 3 using the definition $a - b = x$ if and only if $x + b = a$ for all a, b in Z_7.

Example

$5 = \underline{\hspace{1.5cm}} + 2$

6. Use the $(Z_7, -)$ table to correctly complete the following open sentences.
   a. $6 - 4 = \underline{\hspace{1.5cm}}$, $4 - 6 = \underline{\hspace{1.5cm}}$
   b. $2 - 5 = \underline{\hspace{1.5cm}}$, $5 - 2 = \underline{\hspace{1.5cm}}$
   c. $0 - 3 = \underline{\hspace{1.5cm}}$, $3 - 0 = \underline{\hspace{1.5cm}}$
   d. $6 - (5 - 3) = \underline{\hspace{1.5cm}}$, $(6 - 5) - 3 = \underline{\hspace{1.5cm}}$
   e. $(6 - 3) - 1 = \underline{\hspace{1.5cm}}$, $6 - (3 - 1) = \underline{\hspace{1.5cm}}$

7. If possible, find a, b, and/or c that make the following sentences true.
   a. $6 - a = 6$
   b. $a - 3 = a$
   c. $((a - b) - c) - a = 3$
   d. $a - b = a - c$

2.15 DIVISION IN VARIOUS SYSTEMS

Consider the processes involved in making a definition. Definitions do not suddenly appear but are written to make life easier—to do a job that needs to be done. You defined subtraction in Z_7, because it looked as if subtraction would be useful in Z_7 for solving equations.

In addition, definitions are usually based on prior experience. When you defined subtraction in Z_7, you had a good idea of how subtraction operates in familiar sets like integers or rational numbers.

Finally, it is often necessary to place restrictions on a definition depending on the system in which you work. For example, in W, subtraction of two elements $(a - b)$ can only be defined if a is greater than or equal to b.

Keep these three ideas in mind as you proceed to define division in Z_7.

PROBLEM 1   Hopefully, subtraction has gone along smoothly. You should now be able to deal with division in Z_7. Again think of what you want division to do. Start by trying the following examples in Z_7:

$6 \div 2 =$ _____
$2 \div 4 =$ _____
$5 \div 1 =$ _____
$0 \div 2 =$ _____
$3 \div 0 =$ _____

Definition of Division in Z_7: _____

Once you have a definition, please contact your instructor.

PROBLEM 2   Does your definition of division work for all numbers in Z_7?

Are there any restrictions on the numbers? If so, what are they?

PROBLEM 3   Does your definition of division work in Q?
　　　　　There are many possibilities for the definition of division in Z_7. One would be as follows:

$$a \div b = a \cdot b^{-1} \text{ for all } a, b \text{ in } Z_7, b \neq 0$$

PROBLEM 4   Try using this definition on some division problems in Z_7.
　　a.　$2 \div 5 =$ _____
　　b.　$4 \div 2 =$ _____
　　c.　$3 \div 6 =$ _____
　　d.　$0 \div 2 =$ _____
　　e.　$4 \div 1 =$ _____
　　f.　Why is the restriction $b \neq 0$ needed? _____
　　g.　Can this definition of division be used for all a, b in Q if $b \neq 0$?

Another possibility for the definition of division in Z_7 is:

for all a, b in Z_7, $b \neq 0$, $a \div b = x$ if and only if $x \cdot b = a$

PROBLEM 5   Try using this definition on some division problems in Z_7.
　　a.　$5 \div 3 =$ _____
　　b.　$3 \div 5 =$ _____
　　c.　$0 \div 5 =$ _____
　　d.　$1 \div 2 =$ _____
　　e.　$2 \div 4 =$ _____

f.   Why is the restriction $b \neq 0$ needed this time?_____

g.   Can you apply this definition of division in Q?_____

PROBLEM 6   Now look at W and Z:

a.   Does either of the two definitions of division given in Problems 11 and 12 work for all elements in W?

b.   in Z?

PROBLEM 7   If a and b are elements of W, what restrictions must be placed on a and b so that $(a \div b)$ is an element of W?

PROBLEM 8   If a and b are elements of Z, what restrictions must be placed on a and b so that $(a \div b)$ is an element of Z?

PROBLEM 9   Now check the operations of subtraction and division in the sets Q and Z_7 and see which properties are true and which ones are false. Check closure, commutativity, and associativity, using Table 2.13 to record your results. For each property, determine whether the property is true or false in Q and in Z_7. If a property is true in a given system, write T. However, if a property is false in the given system, write F and give one numerical counterexample in the chart.

Note: Each property needs to be translated into the appropriate system. For instance, suppose you wanted to check commutativity in $(Q, \div)$. The statement of commutativity would be: Division is COMMUTATIVE in Q: For all a, b in Q with a, $b \neq 0$, $a \div b = b \div a$.

TABLE 2.13

PROPERTY	Q	Z_7
Closure under $-$		
Closure under $\div$ by nonzero elements		
Commutativity for $-$		
Commutativity for $\div$		
Associativity for $-$		
Associativity for $\div$		

PROBLEM 10   Filling in Table 2.14 may help you to focus on the relationship between closure, the solution of equations, and the inverse properties. Fill in the table by simply checking the appropriate spot if the property is true.

TABLE 2.14

PROPERTY	W	Z	Q	Z_7
$x + a = b$ has a solution for all a, b				
Additive Inverse				
Closure under $\div$ by nonzero elements				
$a \cdot x = b$ has a solution for all a, b with $a \neq 0$				
Multiplicative Inverse for nonzero elements				

Your investigation in this section has involved two new operations, subtraction and division, which you actually defined (in more than one way) from the more familiar operations of addition and multiplication. You've considered how subtraction and division then operate on different sets by checking for such properties as closure, commutativity, and associativity. Finally you've seen that the closure property, solution of equations, and inverse properties are equivalent in certain circumstances. In your homework you'll further consider identities, inverses, and distributivity when subtraction and division operate on the rationals.

HOMEWORK EXERCISES

1. For each equation given, determine whether the equation has a solution in the given system and find the solution if it exists.

In W: **a.** $x - 5 = 7$
        b. $4 \cdot x - 3 = 5$
        c. $(x - 3) \div 3 = 2$
        d. $6 - x = 14$

In Z: **a.** $x - 2 = 14$
       b. $5 - x = 10$
       c. $^-3x - 6 = 0$
       d. $x - {}^-5 = {}^-7$

In Q: **a.** $x \div 3 = 5$
       b. $^-3 \cdot x - 4 = 3$

c. $\dfrac{2}{3} \cdot x - 3 = \dfrac{2}{3} \cdot x + 5$

d. $(x - \dfrac{5}{2}) \div \dfrac{2}{3} = \dfrac{3}{4}$

In Z_7: a. $x - 6 = 2$

b. $x \div 4 = 3$

c. $0 \cdot x - 5 = 2$

d. $(x - 4) \div 2 = 1$

Examine $(Q, -)$ for the identity property. Remember that the identity element must be commutative, i.e., there exists an identity element e such that $e - a = a - e = a$, for all a in Q.

a. Does the identity property hold in $(Q, -)$?

b. If there is an identity element, list it.

2. Does the inverse property hold in $(Q, -)$? If so, list the inverses of $\dfrac{-2}{3}$, $5, 0, 1.73$. If not, list an element which does not have an inverse.

3. Examine $(Q, \div)$ for the identity property.

a. Does the identity property hold in $(Q, \div)$?

b. If true, what is the identity element?

4. Does the inverse property hold in $(Q, \div)$? If so, list the inverses of $\dfrac{-2}{3}$, $t, 0, 1.73$. If not, list an element which does not have an inverse.

5. Check to see if division is distributive over subtraction in Q.

a. First check right distributivity:

$$(a - b) \div c = (a \div c) - (b \div c) \text{ for } a, b, c \text{ in } Q, c \neq 0$$

b. Now left distributivity:

$$c \div (a - b) = (c \div a) - (c \div b) \text{ for } a, b, c \text{ in } Q, a \neq b$$

6. Check to see if division is distributive over subtraction in Z_7.

2.16 PROPERTIES OF $(Z_n, +, \cdot)$

In a previous section, you have studied properties of the system $(Z_7, +, \cdot)$. In this section you will consider a number of other modular systems.

DEFINITION: For each integer $n \geqslant 2$, let $Z_n = \{0, 1, \ldots, n-1\}$. From this definition, $Z_2 = \{0, 1\}$.

PROBLEM 1   List all the members for each of the following sets.

    a. $Z_3 = \{$ $\}$

    b. $Z_4 = \{$ $\}$

    c. $Z_5 = \{$ $\}$

    d. $Z_6 = \{$ $\}$

For each set Z_n with $n \geqslant 2$, there are operations of addition modulo n and multiplication modulo n.

In the following problems, you will consider the systems Z_2, Z_3, Z_5, Z_6. The goal in these problems is to decide how the mod operations work, to complete the mod addition and mod multiplication tables, and to find additive and multiplicative inverses whenever possible.

Solve all problems before recording your results.

PROBLEM 2   Complete the following tables:

TABLE 2.15
$(Z_2, +)$

+	0	1
0		
1		

$^-0 =$
$^-1 =$

TABLE 2.16
$(Z_2, \cdot)$

$\cdot$	0	1
0		
1		

$1^{-1} =$

TABLE 2.17
$(Z_3, +)$

+		

$^-0 =$
$^-1 =$
$^-2 =$

TABLE 2.18
$(Z_3, \cdot)$

$\cdot$		

$1^{-1} =$
$2^{-2} =$

TABLE 2.19 $(Z_4, +)$

$^-0 =$
$^-1 =$
$^-2 =$
$^-3 =$

TABLE 2.20 $(Z_4, \cdot)$

$1^{-1} =$
$2^{-1} =$
$3^{-1} =$

TABLE 2.21 $(Z_5, +)$

+					

$^-0 =$
$^-1 =$
$^-2 =$
$^-3 =$
$^-4 =$

TABLE 2.22 $(Z_5, \cdot)$

$\cdot$					

$1^{-1} =$
$2^{-1} =$
$3^{-1} =$
$4^{-1} =$

TABLE 2.23 $(Z_6, +)$

+						

$^-0 =$
$^-1 =$
$^-2 =$
$^-3 =$
$^-4 =$
$^-5 =$

TABLE 2.24 $(Z_6, \cdot)$

$\cdot$						

$1^{-1} =$
$2^{-1} =$
$3^{-1} =$
$4^{-1} =$
$5^{-1} =$

You have now constructed addition and multiplication tables for the systems $(Z_n, +, \cdot)$ for several small values of n. The objective is now to determine *all* values of n for which the system $(Z_n, +, \cdot)$ is a field. It is necessary to determine for which values of n, $(Z_n, +, \cdot)$ is a field. One way to do this is to organize the data in a table listing all the field properties. This is a major organizational task.

PROBLEM 3   In Table 2.25 first determine whether the property is satisfied in each of the systems Z_n. If a property is satisfied in a given system, place a check mark in the appropriate spot in the chart. If a property is not satisfied in the given system, write a numerical counter-example in the table (that is, one instance for which the property fails). In the last line

TABLE 2.25

	Z_2	Z_3	Z_4	Z_5	Z_6	Z_7	Z_8	Z_9	*True for all Z_n*
ADDITION Closure									
Associativity									
Additive Identity									
Additive Inverse									
Commutativity									
MULTIPLICATION Closure									
Associativity									
Multiplicative Identity									
Multiplicative Inverse									
Commutativity									
DISTRIBUTIVE LAW									
Field									

Note: Associativity can be checked by means of a computer program. Ask your instructor if such a program is available.

place a check mark for each Z_n examined which is a field. You will probably note that certain properties are the key ones to check.

After completing Table 2.25, if a property appears to hold for all values of n, write Yes in the last column. However, if a property fails for some values of n, write No in the last column and list the values of n for which it fails.

PROBLEM 4   Summarize your results by correctly completing the following statements:

If n is _____ , the system $(Z_n, +, \cdot)$ is a field.

If n is _____ , the system $(Z_n, +, \cdot)$ is not a field.

Have you proved the preceding statements, or are these statements conjectures about $(Z_n, +, \cdot)$?

What was the value of checking Z_8 and Z_9?

HOMEWORK EXERCISES

1. Consider the following statement: If $a \neq 0$ and $b \neq 0$ then $a \cdot b \neq 0$, for all a, b in Z_n.
    a. For which of the modular systems, Z_2 through Z_9, is the statement true? Provide a counterexample for each system in which the statement is false.
    b. Correctly complete the following proposed generalization or conjecture: The statement "If $a \neq 0$ and $b \neq 0$ then $a \cdot b \neq 0$ for all a, b in Z_n" is true if n is _____ and false if n is _____ _____ .

2. Consider the linear equations $ax + b = c$, where a, b, c in Z_n and $a \neq 0$.
    a. For which values of n will all such linear equations have solutions in Z_n? Why?
    b. The general solution (when it exists) of the linear equation $ax + b = c$ in Z_n is given by $x = $ _____ .
    c. Whenever possible, solve the given linear equation in the indicated system.
        $2x + 4 = 3$ in Z_5
        $3x + 5 = 3$ in Z_6
        $5x + 4 = 2$ in Z_6
        $7x + 6 = 1$ in Z_8
        $6x + 7 = 2$ in Z_8
        $10x + 4 = 8$ in Z_{11}

3. Solve each of the following equations in 2 ways:
   a. by use of the general solution found in Problem 2b.
   b. by substituting every element in the system into the equation.

   $6x - 5 = 7$ in Z_9
   $5x + 2 = 4$ in Z_8
   $2 + 3x = 5$ in Z_6
   $2x - 4 = 6$ in Z_{12}

4. If a modular system is a field, *every* linear equation with nonzero coefficient of x has *exactly* one solution because every element has a multiplicative inverse. However, this is not the case in nonfields.
   a. How can you tell whether or not an element will have a multiplicative inverse in a modular system that is not a field?
   b. What condition guarantees a solution to the equation $ax + b = c$ in a modular system that is not a field?

5. Fill in the blanks as required, and prove all of the following statements about Z_5:
   a. The smallest counting number k such that $a^k = 1$ for all $a \neq 0$ in Z_5 is $k = \underline{\hspace{1cm}}$, k in N.
   b. $a^5 = \underline{\hspace{1cm}}$ for all a in Z_5.
   c. $(a + b)^5 = a + b$ for all a, b in Z_5.
   d. $(a + b)^5 = a^5 + b^5$ for all a, b in Z_5.

6. Examine the results of Exercise 5 about Z_5 and similar results about Z_7 (see Section 2.6). Write four comparable statements about Z_4, and determine the truth or falsity of these statements.

★7. Generalize the results of Exercises 5 and 6 by determining *all values of n* for which the following conjectures are true:
   a. $a^{n-1} = 1$ for all $a \neq 0$ in Z_n.
   b. $a^n = a$ for all a in Z_n.
   c. $(a + b)^n = a + b$ for all a, b in Z_n.
   d. $(a + b)^n = a^n + b^n$ for all a, b in Z_n.

SUMMARY

In Chapter 2 you have examined a number of different systems and how they are structured by binary operations. There are a number of ways to organize the subject matter of mathematics.

Certainly using the concept of binary operation is a good one—binary operations are very common from the most elementary mathematics to the most advanced.

Much time was spent organizing Z_7, an unfamiliar set of numbers. New insight into structure was provided by working in Z_7 and other finite number systems. The familiar binary operations of addition and multiplication created the basic structure of a field. The inverse operations of subtraction and division were explored and defined. All of this provides a basis for our work in whole numbers, integers, and rational numbers.

Continuously, the processes of mathematics were emphasized: organizing data, exploring patterns, formulating conjectures and definitions, determining conditions for statements and definitions, testing or confirming statements, organizing statements, proving statements, comparing systems, testing systems. This interplay of inductive and deductive processes is the heart of working in mathematics. In Section 2.16 this is clearly illustrated as an interesting theorem in field theory is developed.

Binary Operations and Groups

3

3.1 NONSTANDARD OPERATIONS AND THEIR PROPERTIES

The family of modular arithmetic systems, $(Z_n, +, \cdot)$, was produced by focusing on special sets of numbers and new ways of adding, subtracting, multiplying, and dividing. This modular arithmetic is helpful in solving problems ranging from the calendar to secret codes. The modular arithmetic operations share many properties with standard arithmetic in W, Z, and Q. These are properties that are often very helpful in solving problems.

In this chapter you will investigate an even broader collection of operations—some strange, many important, all with interesting properties. Some of these new operations will not even be related to the familiar operations of addition, subtraction, multiplication, and division.

PROBLEM 1 Following are the results of applying an operation # (called sharp) to several pairs of rational numbers. Study the examples and then complete the others according to the rule you sense in the examples.
  a. $^-4 \, \# \, 7 = 7$
  b. $12 \, \# \, 25 = 25$
  c. $1.3 \, \# \, ^-6.1 = 1.3$
  d. $12 \, \# \, 9 = 12$

e. $15 \# 14 =$ _____

f. $^-3 \#^-5 =$ _____

g. $0 \# 21 =$ _____

h. $\dfrac{5}{3} \#$ _____ $= \dfrac{7}{2}$

i. $7 \#$ _____ $= 7$

j. _____ $\#^-16 = {}^-16$

k. $33 \# 33 =$ _____

l. $91 \#$ _____ $= 91$

m. _____ $\# 3 = 14.6$

n. $(4 \#^-3) \# 8 =$ _____

o. $4 \# (^-3 \# 8) =$ _____

p. $6 \#$ _____ $= 6$

q. (_____ $\# 3) \#^-8 = 13$

r. _____ $\# (3 \#^-8) = 13$

s. $\dfrac{21}{43} \# \dfrac{8}{17} =$ _____

t. $3.548 \# 3.546 =$ _____

u. State a rule for $a \# b =$ _____

To investigate some of the basic properties of #, judge each of the statements in Problems 2 to 5 as true or false, in (W, #), (Q, #) and (P, #). Let $P = \{5, 10, 15, 20, \ldots,\}$.

PROBLEM 2   For any numbers s, t, $s \# t = t \# s$. (Commutative property for #.) Check in W, then Q, then P.

Provide examples if true or counterexample if false.

PROBLEM 3   For any numbers s, t, u, $(s \# t) \# u = s \# (t \# u)$. (Associative property for #.) Check in W, then Q, then P.

Provide examples if true or counterexamples if false.

PROBLEM 4   There is a fixed number e such that $s \# e = e \# s = s$ for all s. (Identity property for #.) Check in W, then Q, then P.

List the identity element for each system in which it exists.

PROBLEM 5   For each element s, there is an element t such that $s \# t = t \# s = e$. (Inverse property for #.) Check in W, then Q, then P.

Provide examples of inverses for each system in which they exist.

Think a moment about the inductive and deductive cycle you have used in this section. Inductively you have explored several given examples of the operation # in order to discover a rule that fits all the examples. Once you defined the operation by the rule, you tested it in different sets to determine those field properties which hold and those which fail.

HOMEWORK EXERCISES

1. For each of the following equations in $(Z, \#)$, find all possible solutions or show that none exist.
   a. $p \# 3 = 12$
   b. $(p \# 3) \# p = 3$
   c. $(7 \# 6) \# p = 3$
   d. $p \# 8 = q \# 8$
   e. $8 \# p = 8$
   f. $(3 \# 4) \# p = p \# 7$
   g. $(^-8 \# ^-3) \# p = ^-2$
   h. $p \# (^-10 \# ^-14) = ^-12$

2. Which of the following sets of numbers are *closed* under the operation #? Explain your answers.
   a. W
   b. Z
   c. Q

3. Let + and · stand for the ordinary whole number operations. Determine which of the following statements involving a mix of #, +, and · are true and which are false. Try to give a justification for those that are true and give counterexamples for those that are false.
   a. For all p, q, r in W, $p \# (q + r) = (p \# q) + (p \# r)$.
      (# distributing over +)
   b. For all p, q, r in W, $p + (q \# r) = (p + q) \# (p + r)$.
      (+ distributing over #)
   c. For all p, q, r in W, $p \cdot (q \# r) = (p \cdot q) \# (p \cdot r)$.
      (· distributing over #)

4. Complete Table 3.1 for # on the first six elements of W.

TABLE 3.1

#	0	1	2	3	4	5
0						
1						
2				4		
3						
4						
5						

Problem 1 involved the exploration of one of an almost limitless supply of different operations using an assignment process fitting the following basic form:

DEFINITION: If S is a set of numbers and x and y represent elements of S, a *binary operation*, $*$, on S assigns to each ordered pair of numbers (x, y) exactly one number which is denoted by $z = x * y$. S is said to be *closed under* $*$ if and only if $(x * y)$ is in S for all x, y in S.

PROBLEM 6 Complete the following statements using the symbol, $*$, (star) to represent any binary operation and S to represent any set.

a. $(S, *)$ has an identity element e if and only if $x * \underline{\hspace{1cm}} = \underline{\hspace{1cm}} * x$ $= x$ for all x in S.

b. A number x has a $*$ *inverse* in S, (denoted x') if and only if $x * x'$ $= x' * x = \underline{\hspace{1cm}}$.

c. $*$ is associative on S if and only if $\underline{\hspace{3cm}}$.

d. $*$ is commutative on S if and only if $\underline{\hspace{3cm}}$.

PROBLEM 7 The next examples illustrate another, probably familiar, operation on numbers. Study the given examples and complete the rest.

a. $7 \gamma 13 = 10$

b. $12 \gamma 18 = 15$

c. $^-6 \gamma 4 = ^-1$

d. $6 \gamma 5 = 5\frac{1}{2}$

e. $8 \gamma 8 = 8$

f. $^-18 \gamma ^-12 = \underline{\hspace{1cm}}$

g. $5 \gamma 8 = $ _____

h. _____ $\gamma\ 23 = 25$

i. _____ $\gamma\ 23 = 5$

j. $(6\ \gamma\ 8)\ \gamma\ 12 = $ _____

k. $a\ \gamma\ b = $ _____

l. $6\ \gamma\ (8\ \gamma\ 12) = $ _____

m. $(a\ \gamma\ b)\ \gamma\ c = $ _____

n. $a\ \gamma\ (b\ \gamma\ c) = $ _____

PROBLEM 8 Investigate the basic properties of $(Q,\ \gamma)$. The basic properties are: closure, associativity, identity, inverse, and commutativity.

PROBLEM 9 Investigate the basic properties of $(Z,\ \gamma)$.

HOMEWORK EXERCISES

5. Solve each of the following equations in $(Q,\ \gamma)$.
    a. $x\ \gamma\ 7 = 12$
    b. $(7\ \gamma\ x)\ \gamma\ x = 25$
    c. $x\ \gamma\ 10 = x$
    d. $8\ \gamma\ x = 8\ \gamma\ y$

6. If + and · represent ordinary addition and multiplication, check to see whether:
    a. + distributes over γ: $p + (q\ \gamma\ r) = (p + q)\ \gamma\ (p + r)$.
    b. · distributes over γ.

7. Is it possible to calculate the average of three numbers p, q, r by use of the γ operation? Illustrate and explain your answer.

★8. Complete Table 3.2 for $a\ \gamma\ b = (a + b) \div 2$ in $(Z_5,\ +,\ \cdot)$.

    Note: The division $(a + b) \div 2$ in mod 5 means $(a + b) \cdot 2^{-1}$ in mod 5.

TABLE 3.2

γ	0	1	2	3	4
0					
1					
2					
3				1	
4					

★**9.**   Check the basic properties of $(Z_5, \overset{\bullet}{\gamma})$.

PROBLEM 10   Take turns in your group leading the following contest:
   a. Pick a set and privately choose a rule for an operation on that set.
   b. Put several examples on the board to help other group members guess what you're thinking of. For instance:

$$3 * 5 = 10$$
$$6 * 4 = 20$$
$$3 * 8 = ?$$

   c. If the group can't guess, give another hint.

$$(3 * 8) = 16$$

   d. The winner is the first who reads your mind by deciphering the pattern in examples.

Did you guess $a * b = (a - 1) \cdot b$?

To define a binary operation you must show how the operation acts on any two elements. The order of the elements is important if the operation is not commutative. This defining procedure can be accomplished in three different ways: presenting a table, stating a rule, or listing the assignment of all *ordered pairs*.

The familiar operation of addition in Z_3 can serve as an example.

TABLE 3.3

+	0	1	2
0	0	1	2
1	1	2	0
2	2	0	1

RULE

$$a + b = \begin{cases} a + b \text{ if } a + b < 3, \, a, \, b \in W \\ a + b - 3 \text{ if } a + b \geqslant 3, \, a, \, b \in W \end{cases}$$

ASSIGNING ORDERED PAIRS

$$(0,0) \xrightarrow{+} 0$$
$$(0,1) \xrightarrow{+} 1$$
$$(0,2) \xrightarrow{+} 2$$
$$(1,0) \xrightarrow{+} 1$$
$$(1,1) \xrightarrow{+} 2$$
$$(1,2) \xrightarrow{+} 0$$
$$(2,0) \xrightarrow{+} 2$$
$$(2,1) \xrightarrow{+} 0$$
$$(2,2) \xrightarrow{+} 1$$

PROBLEM 11   An operation, $, can be defined on the set N by the following rule: $a \$ b = a^b$ for all a, b in N. Take the first three elements of N and show how you would define $ in terms of

a.   a table
b.   an assignment of ordered pairs

Of the three ways of defining an operation what is the advantage of a rule when you are considering the operation on an infinite set?

HOMEWORK EXERCISES

10.   Operations can be defined directly on sets of numbers by stating rules. For example,
$a \bigtriangledown b = a$ for all a, b in Z
$a \text{ m } b = $ smaller of a and b in Q
$a \wedge b = 3$ for all a, b in W

Using the preceding rules, calculate:
a. $5 \bigtriangledown 11 =$
b. $11 \bigtriangledown 5 =$
c. $^-64 \text{ m } ^-52 =$
d. $\dfrac{5}{36} \text{ m } \dfrac{1}{7} =$
e. $5 \wedge 12 =$
f. $5 \wedge (7 \bigtriangledown 8) =$
g. $(5 \wedge 7) \bigtriangledown 8 =$

★11. One of the basic purposes of studying nonstandard binary operations is to highlight the essential structure of familiar, more practical systems. As a teacher, you might often find it helpful to use examples of non-commutative operations to clarify the precious virtue of commutativity in systems like $(W, +)$ or $(Q, \cdot)$. For each of the following combinations of properties list several systems (set, operation) that fit the conditions.

a. commutative, not associative
b. associative, not commutative
c. identity, inverse, associative
d. commutative, identity, not associative

   Hint: None of the previous operations satisfies this set of conditions.

★12. Find two operations on W in the list in Homework Exercise 10 such that one distributes over the other. Write the meaning of the property in terms of the symbols for the operations chosen. Illustrate the distributivity with three different triples of numbers.

3.2 OPERATIONS ON POINTS OF A PLANE

Until now, binary operations have been considered only on sets of numbers. These binary operations have assigned a single number as the sum, product, or maximum for given pairs of numbers. Similarly, an operation may be defined on the points of a plane. Some very attractive geometric patterns arise, and algebraic properties take on a visual flair.

The definition of a binary operation in Section 3.1 refers only to sets of numbers. This definition can be easily generalized to include more than just operations on numbers.

DEFINITION: Let S be a set. A *binary operation* on the set S assigns to each ordered pair of elements from S exactly one element which may or may not be an element of S.

You are encouraged to compare the definition to the one given previously to convince yourself that both definitions actually convey the same concept.

Let S be the set of all points in a plane and let *mid* be the operation which assigns each pair of points of the plane (A, B) a point labeled A mid B as shown in Figure 3.1.

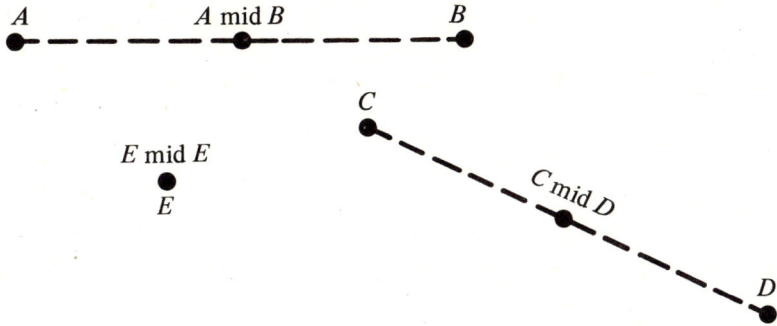

FIGURE 3.1

PROBLEM 1   Investigate the basic properties of this new operation, mid, on the set of points of the plane of the blackboard. Remember the basic properties are closure, associativity, identity, inverse, commutativity.

PROBLEM 2   Locate two points M and N on the chalkboard.
  a. Find all points X such that X mid $M = N$
      (**Note:** All such points X are solutions of the equation X mid $M = N$.)
  b. Find all solutions of Y mid $N = M$.
  c. Find all solutions of N mid $Z = M$.

PROBLEM 3   Find the set of all points that can be obtained from two *fixed points, P* and Q, under the mid operation applied many times. First apply mid to P and Q, then to P and $(P$ mid $Q)$, then to Q and $(P$ mid $Q)$; continue in this fashion, using any two of the new points found.

HOMEWORK EXERCISES

There are many other examples of binary operations on points on a plane. A few more will be examined.

1. The operation Right Turn, ⅂, assigns a *point* in the plane to any given

pair of points according to the rule illustrated in Figure 3.2. Notice that the motion is clockwise. $A \text{ } ⅂↓ \text{ } B = C$ where A, B, C, D are vertices of a square. **Note** that $C \text{ } ⅂↓ D = A$ with this rule.

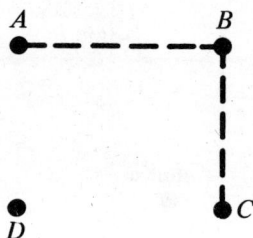

FIGURE 3.2

a. What point is $B \text{ } ⅂↓ A$?
    (**Hint:** This point is not the same as $A \text{ } ⅂↓ B$.)
b. Find the point X where $X = A \text{ } ⅂↓ (A \text{ } ⅂↓ B)$?
    (**Note:** $A \text{ } ⅂↓ (A \text{ } ⅂↓ B) = A \text{ } ⅂↓ C$.)
c. Check the basic properties of this "squaring" operation.
d. Starting from two fixed points, P and Q, what points can you generate by repeatedly using $⅂↓$?

2. The operation, tri, assigns to points A and B the point C such that A, B, and C determine an equilateral triangle.
    a. Sketch A tri B below.

FIGURE 3.3

b. Did you hesitate over where to locate the new point? The rule is ambiguous! Add some clarifying restrictions to the rule.
c. Now use this precise tri operation for checking the basic properties.

★3. The set on which a binary operation can be defined need not consist solely of numbers or points on a plane. The next exercise calls for imagination in setting up binary operations.

Describe and illustrate binary operations that assign
a.   A *line* to any given pair of *lines* in the plane.
b.   A *segment* (•————•) to any given pair of *segments* in the plane.
c.   An *angle* to any given pair of *angles* in the plane.
Note: Remember that the definition of binary operation stipulates the assignment of *exactly one* element to the ordered pair of elements.

★**4.** Check the basic properties for each operation you devised in Exercise 3.

Now you've seen how binary operations can be defined on geometric sets such as points, line segments, lines, and angles. Once defined, you've tested the binary operation on geometric sets for properties like closure, inverse, identity, commutativity, and associativity in exactly the same way you did for sets of numbers. The concept of binary operations is an important one to structure a variety of different sets in mathematics.

3.3   INTRODUCTION TO SETS AND SET NOTATION

We shall think of a set as a collection of well-defined elements. When we say elements belonging to a set are "well-defined," we mean that in order for a given collection to be considered a mathematical set we must be able to determine whether any object belongs or does not belong to the collection. If the identification of what belongs or what does not belong to the collection can be made, then the collection is a mathematical *set,* otherwise it is not.

There are several ways to denote sets. A set can be denoted by listing the elements of the set between braces, $\{\ \}$. For example, $A = \{0, 1, 2, 3, 4, 5\}$ denotes the set of whole numbers which are less than or equal to 5.

Note: It is customary to denote sets with capital letters. Also, to indicate that 2 is an element of the set A, we would write $2 \in A$.

Another way of denoting a set is to use some identifying property of the elements of the set, for example, an equation or an inequality with proper notation, and to enclose this within braces—called set builder notation. For example, $\{x \mid x + 2 = 5,$ $x \in W\}$ denotes the set of all x such that $x + 2 = 5$ and x is a whole number.

Note: The | is read "such that."

If two sets, A and B, are related in such a way that every element belonging to A is also an element of B, then A is defined to be a subset of B. This is denoted $A \subseteq B$. If A is not a subset of B, then A would contain some elements not in B.

We say that two sets, A and B, are *equal* if and only if A and B contain the same elements.

PROBLEM 1   If $A \subseteq B$ and $B \subseteq A$, what other relation exists between A and B? Explain.

If we wished to identify the set of Republican voters in the United States, we would select those elements from the universal set of all voters in the United States. In our work with sets, we will use the symbol U to denote the universal set which we use in talking about other sets.

Suppose that A is a set chosen from the universal set U. A common way to describe this situation pictorially is with a diagram such as Figure 3.4.

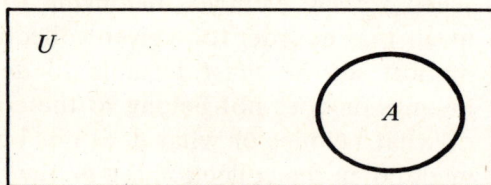

FIGURE 3.4

The set consisting of all elements in U which are not in A is called the *complement* of A and is denoted $\overline{A}$.

PROBLEM 2   Shade the area of the diagram above which corresponds to $\overline{A}$. What is the complement of U, i.e., $\overline{U}$?

We need notation for a set which contains no elements. It is commonly referred to as $\{\}$. The symbol $\emptyset$ is also used to refer to this set. In either case this set is called the *null set* or the *empty set*. Note that the empty set is a subset of every set: if $\{\}$ were not a subset of any given set, then $\{\}$ would contain some element not in the given set. But $\{\}$ has no elements!

Now consider the set of all subsets of U, denoted by $P(U)$. Note that $P(U)$ is a set with elements which are themselves sets. If $U = \{0, 1\}$, the elements of $P(U)$ would be: $\{\}$, $\{0\}$, $\{1\}$, and $\{0, 1\}$. $P(U)$ is read "P of U" which means the power set of the universal set.

PROBLEM 3   Let $U = \{0, 1, 2, 3\}$. List the elements of $P(U)$.

HOMEWORK EXERCISES

1. Which of the following constitute a set according to the preceding definition?
    a.   all the people in your class who are over 19 years old
    b.   all the people in your class who are under 15 years old
    c.   all the bright women in your class
    d.   all the interesting courses offered at the university
    e.   all the even whole numbers
    f.   all the odd numbers that are whole numbers
    g.   all the whole numbers greater than 43

2. Write each of the sets from Homework Exercise 1 in roster (listing each element) *or* set builder (equation or inequality) notation.

3. Consider the descriptions in Homework Exercise 1. Find the complement of each one you determine as a set using the following as the universal set.
    a.   through (c) all the people in your class,
    d.   all the courses offered at the university,
    e.   through (g) all the whole numbers.

4. If $\overline{A} = B$, is it true that $\overline{B} = A$?

5. a.   Indicate the inclusion relations among the following sets of numbers (i.e., which ones are subsets of others).

    $A = \{2, 4, 6, 8, 10\}$

$$B = \{x | 1 < x \leqslant 10, x \in W\}$$
$$C = \{1, 3, 6, 8\}$$
$$D = \{10, 8, 6, 4, 2\}$$
$$E = \{1, 1, 3, 6, 3, 8, 6\}$$

b. Are any of these sets equal?

6. If, for some element x, $x \in A$ and $x \notin B$, is it possible that $A \subseteq B$? $B \subseteq A$? Why or why not?
   Note: "$x \notin B$" is read "x is not an element of B."

★3.4   BINARY OPERATIONS ON $P(U)$

Given a set U and the set $P(U)$ of subsets of U, there are two familiar binary operations which we define on $P(U)$. These are union, $\cup$, and intersection, $\cap$.

We make the following definitions: The *union* of two sets, A and B, from $P(U)$ is the set containing all the elements of A or B.

Note: All elements of A are included and all elements of B are included.

PROBLEM 1   Shade in the portion of Figure 3.5 that corresponds to $A \cup B$.

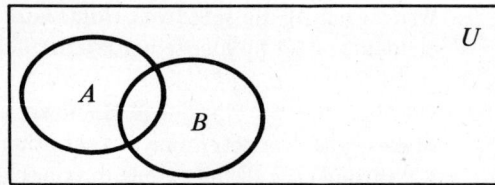

FIGURE 3.5

The *intersection* of two sets, A and B, from $P(U)$ is the set consisting of all elements which A and B have in common.

PROBLEM 2   Shade in the portion of Figure 3.6 that corresponds to $A \cap B$.

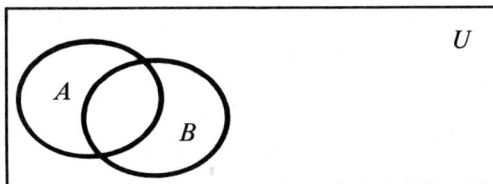

FIGURE 3.6

PROBLEM 3   Let $U = \{0, 1, 2, 3\}$. You found $P(U)$ for this set in Section 3.3 (Problem 3). Now find the set which results from each of the following:

a. $\{0, 1\} \cup \{2, 3\}$

b. $\{0, 1\} \cup \{1, 2\}$

c. $\{0, 1\} \cap \{1, 2, 3\}$

d. $\{0, 1\} \cap \{\overline{2}\}$

e. $\{0, 1\} \cap \{2, 3\}$

You have examined a set and an operation defined on this set before in such systems as $(Z_7, +)$. Now consider the system $(P(X), \cup)$ where $X = \{0, 1\}$ and $\cup$ is the *union* operation on 2 sets.

PROBLEM 4   Complete the following operation table.

TABLE 3.4

$\cup$	$\{\ \}$	$\{0\}$	$\{1\}$	$\{0,1\}$
$\{\ \}$	$\{\ \}$		$\{1\}$	
$\{0\}$				$\{0,1\}$
$\{1\}$		$\{0,1\}$		
$\{0,1\}$				

a. Is $P(X)$ closed under the union operation?
b. Is $P(X)$ commutative under the union operation?
c. Is $P(X)$ associative under the union operation?
d. Is there an identity element for $(P(X), \cup)$? If so, what is it?
e. Are there inverses for each element in $(P(X), \cup)$? If so, list them.

85

PROBLEM 5   Again, let $X = \{0, 1\}$. Using intersection ($\cap$) as the operation instead of union, construct the operation table for $(P(X), \cap)$ and answer the questions in Problem 4 (a-e).

PROBLEM 6   State the distributive law for union over intersection. Use Figure 3.7 to decide whether it is true or false.

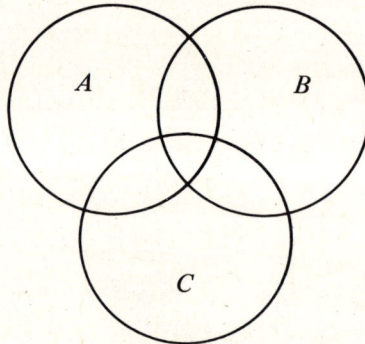

FIGURE 3.7

PROBLEM 7   State the distributive law for intersection over union and decide whether it is true or false.

The definition of a field (see Figure 2.1) is restricted to a set F and two operations, addition and multiplication. In order to investigate the field properties of union and intersection on the power set of X you can define addition and multiplication as follows:

$$A + B = A \cup B$$
$$\qquad \text{for } A \text{ and } B \text{ in } P(X)$$
$$A \cdot B = A \cap B$$

Note that the 0 and 1 elements will be *sets* acting the roles of 0 and 1 according to the definition.

PROBLEM 8   Is $(P(X), \cup, \cap)$ a field? Justify your answer.

So far you have explored binary operations on sets of numbers and geometric sets. Now you have seen how some new binary

86

operations, union and intersection, can be defined on the power sets of given sets. You checked this new structure, union and intersection on power sets, for the various field properties. Your investigation of this structure has proceeded in the same manner as before, but it has revealed its own special characteristics. For example, under the union operation, the empty set takes on the role of the identity element: $A \cup \{ \} = A$ for all A in $P(U)$, and the universal set, U, takes a role similar to 0 under multiplication: $A \cup U = U$ for any set A in $P(U)$. Under intersection, these two sets switch roles with the universal set, U, becoming the identity element: $A \cap U = A$ for all A in $P(U)$ and the empty set acting like 0 under multiplication: $A \cap \{ \} = \{ \}$ for A in $P(U)$.

3.5 MOTIONS OF AN EQUILATERAL TRIANGLE

PROBLEM 1 **a.** From a sheet of paper (heavy stock, if possible) cut out an equilateral triangular region the same size as the one in Figure 3.8.

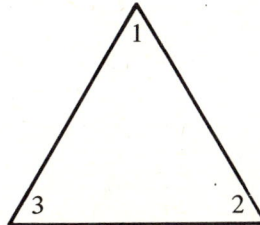

FIGURE 3.8

b. Label the vertices 1, 2, and 3 on both sides of the triangle so that each vertex is labeled with the same number on both sides.

The set to be considered will contain six elements. These six elements will be all of the *rigid motions* of the equilateral triangle. Problems 2 to 7 will describe these motions in detail.

PROBLEM 2 Place your triangle on top of the one on this paper so that the numbers coincide. This position will be called the *initial position*. Rotate your triangle clockwise $\frac{1}{3}$ of a full turn. The vertex numbered 1 on your triangle should coincide with the vertex numbered 2 on this sheet. Figure 3.9 illustrates this motion which will be called R_1. (These turning motions are also referred to as *rotations,* thus the use of R to denote them.)

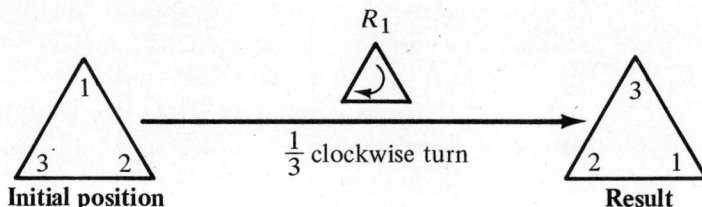

FIGURE 3.9

PROBLEM 3   Starting at the initial position, what other possible rotations can you find? Remember: after each rotation, the triangle should look exactly as it did in the initial position except for the numbers.

PROBLEM 4   You should have found two more rotations. Use the notation in Figure 3.9 to describe (a) R_2: a $\frac{2}{3}$ clockwise turn and, (b) R_3: a $\frac{3}{3}$ (full) clockwise turn.

Check with your instructor to be sure that you understand each of these first three motions.

The remaining three elements of the set of motions can all be classified as flips or *line reflections.*

PROBLEM 5   Starting with your triangle in the initial position, flip the triangle around the vertical line through the top vertex. The bottom left and bottom right vertices of the triangle exchange places. Notice that this motion is the first that requires the triangle to be turned over; it was for this reason that both sides were labeled. Figure 3.10 illustrates the flip around the line through the top vertex, F_T.

PROBLEM 6   There are two more flips. Use the notation in Figure 3.10 to describe:
a. F_R: a flip around the line through the bottom right vertex.
b. F_L: a flip around the line through the bottom left vertex.

PROBLEM 7   a.   Which vertex remains fixed for flip F_R?

88

b. Which vertex remains fixed for flip F_L?

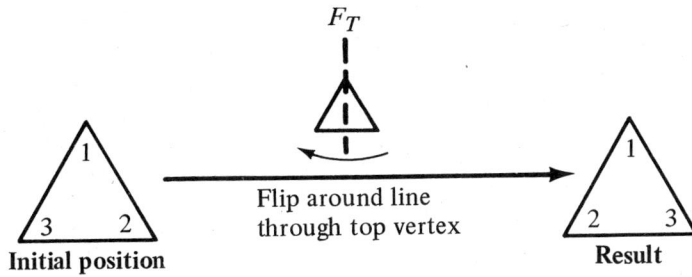

FIGURE 3.10

Again, check with your instructor to make sure that you understand each of the three flips.

PROBLEM 8   As a brief summary of the six motions, and for easy reference, label the vertices of the triangles in Figure 3.11 with the numbers of the results of performing the indicated motions.

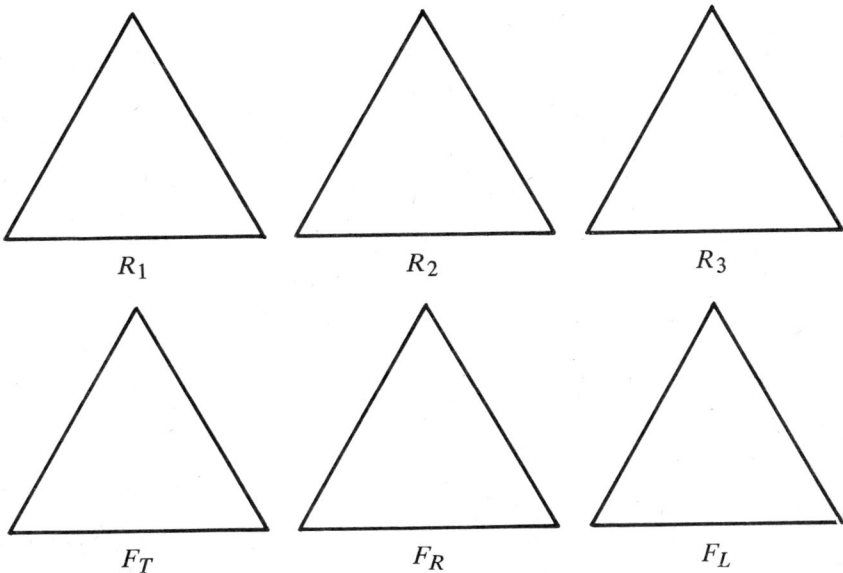

FIGURE 3.11

An operational system, (M, o) can be defined where $M = \{ R_1, R_2, R_3, F_T, F_R, F_L \}$ and o stands for the operation called *composition*. Composition is defined in the following manner:

If a and b represent elements of M, $a \text{ o } b$ is the motion equivalent to performing a from the initial position and then, *from the new position* performing b. ($a \text{ o } b$ is read "a followed by b" or "a composed with b.")

It is important to notice that you do *not* return to the initial position before performing the second motion in a composition.

PROBLEM 9   **a.** To solve the equation $R_1 \text{ o } F_R = X$ for X, begin with your triangle in the initial position and perform R_1 (a $\frac{1}{3}$ clockwise turn). Now, without moving your triangle back to the initial position, perform F_R (a flip around the line through the bottom right vertex).

Compare this result to the triangles in Problem 8. Then $X =$ _____. In the same way, solve:
b. $F_T \text{ o } R_2 = X$ $X =$ _____
c. $R_1 \text{ o } R_2 = X$ $X =$ _____
d. $F_R \text{ o } F_R = X$ $X =$ _____

PROBLEM 10   Complete the following operation table for (M, o).

TABLE 3.5

o	R_1	R_2	R_3	F_T	F_R	F_L
R_1		R_3			F_L	
R_2						
R_3						
F_T		F_R				
F_R					R_3	
F_L						

PROBLEM 11   Is (M, o) closed? Justify your answer.

PROBLEM 12   Does (M, o) have an identity element? If so, list it.

PROBLEM 13   If you found an identity element, list the inverse of each element that has one.

PROBLEM 14   Is (M, o) associative? Since it is too time consuming to check all examples without the use of a computer, check several cases.

PROBLEM 15   If possible, solve the following equations in (M, o).
  a. $X o F_T = R_1$ $X =$ _____
  b. $F_L o Y = R_2$ $Y =$ _____

PROBLEM 16   Using your table for (M, o) and *without* actually performing any motions with your triangle, solve the following where R_1' stands for the inverse of R_1.
  a. $R_1' o F_L' = X$ $X =$ _____
  b. $F_R' o F_T' = X$ $X =$ _____
  c. $F_T' o R_2' = X$ $X =$ _____

PROBLEM 17   **a.** Take a 3-by-5 inch note card and determine the set of motions for a rectangle.
  b. Call the set of motions X and construct an operation table for (X, o).
  c. List the identity element for (X, o) and list the inverse of each element.

HOMEWORK EXERCISES

1. **a.** Cut a piece of paper that fits exactly on top of the pictured triangle.

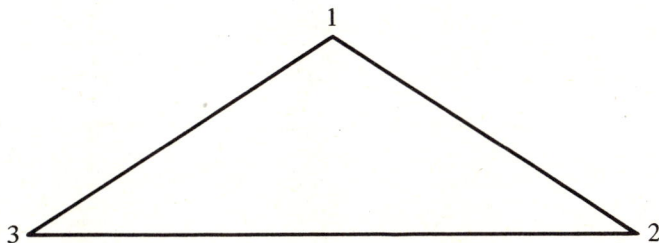

FIGURE 3.12

  b. How many rigid motions can you perform on this triangle? (Remember that the triangle in the new position must fit precisely on top of the original triangle.)

c. Call the set of rigid motions V, and with composition as defined earlier, make a table for (V, o).

d. Investigate the basic properties of (V, o).

2. A system of rigid motions can be developed for some of the following figures: Consider only those systems with more than the identity element.

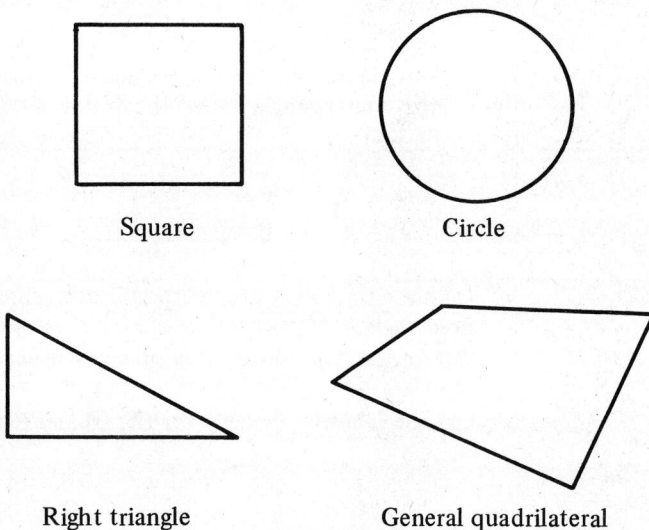

Square   Circle

Right triangle   General quadrilateral

FIGURE 3.13

a. Decide which figures have such a system of rigid motions.

b. List the rigid motions for each figure with such a system.

Those figures for which such motions, other than the identity, exist are said to be *symmetric*.

3.6   A BASIC DEFINITION AND SOME NOTATIONS

Earlier you encountered the system $(Z_7, +)$ and then looked at the properties of modular addition for Z_2, Z_3, Z_4, Z_5, Z_6.

PROBLEM 1   Look back at your findings on properties of the systems $(Z_4, +)$, $(Z_7, +)$, $(X, \circ)$, and $(M, \circ)$. Using this information, complete the following chart. (Use a check or write yes to indicate that the property holds in the named system.)

TABLE 3.6

Property	$(X, \circ)$ Rectangle	$(M, \circ)$ Equilateral Triangle	$(Z_4, +)$	$(Z_7, +)$
Closure				
Associativity				
Identity				
Inverse				
Commutativity				

PROBLEM 2   List those properties in Table 3.6 that hold for all four systems.

DEFINITION: From now on the properties from problem 2 will define a *group*. In other words, a group will be any system, $(G, *)$, (a set, with an operation defined on that set) which has the following properties: Closure, associativity, identity, inverse (every element in the system must have an inverse).

When, in addition to the above, a group is also commutative, it will be known as an *Abelian* or *commutative group*.

Following is the table for the set of motions of a square under the operation composition. The set of motions, *S,* is made up of the following motions:

I = a full turn

R_1 = a clockwise $\frac{1}{4}$ turn

R_2 = a clockwise $\frac{1}{2}$ turn

R_3 = a clockwise $\frac{3}{4}$ turn

H = a flip over the horizontal axis

V = a flip over the vertical axis

D_1 = a flip over the line from the upper right to the lower left vertex

D_2 = a flip over the line from the upper left to the lower right vertex

TABLE 3.7

o	I	R_1	R_2	R_3	H	V	D_1	D_2
I	I	R_1	R_2	R_3	H	V	D_1	D_2
R_1	R_1	R_2	R_3	I	D_1	D_2	V	H
R_2	R_2	R_3	I	R_1	V	H	D_2	D_1
R_3	R_3	I	R_1	R_2	D_2	D_1	H	V
H	H	D_2	V	D_1	I	R_2	R_3	R_1
V	V	D_1	H	D_2	R_2	I	R_1	R_3
D_1	D_1	H	D_2	V	R_1	R_3	I	R_2
D_2	D_2	V	D_1	H	R_3	R_1	R_2	I

PROBLEM 3 a. Check Table 3.7 for the five properties mentioned in connection with a group.
b. Is (S, o) a group? Justify your answer.
c. Is it a commutative group? Justify your answer.

A group provides a way of organizing the properties of a single binary operation defined on a set. The properties used to define a group have been shared by many of the binary operations on sets that you have explored. In Section 2.9 you investigated various systems with respect to the field properties; now this same type of investigation is being done with respect to group properties.

PROBLEM 4 What are the essential differences between a group and a field?

PROBLEM 5 a. How many identities are there in a field? In a group?
b. How many *types* of inverses are there in a field? In a group?

PROBLEM 6 What field property can *never* be discussed with respect to a group? Why?

PROBLEM 7   Decide whether the following is true or false and justify your answer.

If $(F, +, \cdot)$ is a field, then $(F, +)$ and $(F, \cdot)$ are both commutative groups. Give a specific example or counterexample.

★PROBLEM 8   Is the following statement true or false? Justify your answer.

If $(G, *_1)$ and $(G, *_2)$ are both commutative groups, then $(G, *_1, *_2)$ is a field.

Note: In this problem + in a field is defined as $*_1$; $\cdot$ in a field is defined as $*_2$.

You have been using various notations for identities and inverses in several systems. For addition, the identity is called 0 and the inverse is denoted ^-a. In multiplication, the identity is 1 and the inverse is denoted a^{-1}. Finally in the general group, G, with operation $*$, the identity will be designated by e and the inverse of a will be denoted by a'.

TABLE 3.8

Operation	Identity	Inverse of a
+	0	^-a
.	1	a^{-1}
*	e	a'

HOMEWORK EXERCISES

1. **a.** $a * b'$ is an expression which uses the notation for a general group. Write the corresponding expression using the notation for addition. Write the corresponding expression using the notation for multiplication.
    b. $^-a + b$ is written in the notation for addition. Write the corresponding expression for the general group.
    c. $a^{-1} \cdot b$ is written in the notation for multiplication. Write the corresponding expression for the general group.

2. $Z_3 - \{0\}$ denotes the set $\{1, 2\}$. $(Z_3 - \{0\}, \cdot)$ is the system $(\{1, 2\}, \cdot)$ where multiplication is Z_3 multiplication. Determine whether $(Z_3 -$

$\{0\}$, $\cdot$) and $Z_6 - \{0\}$, $\cdot$) are groups, and if so are they commutative (Abelian) groups?

3. Is $(Z_3, \cdot)$ a group? Why or why not?

4. Test as many other systems $(Z_n - \{0\}, \cdot)$ as necessary to complete the following conjecture. The system $(Z_n - \{0\}, \cdot)$ is a group if and only if n is _____ .

5. On the basis of your previous work in fields, decide whether $(Z_n, +)$ is a group for all values of n.

6. Write a formal definition of a general group $(G, *)$ using the notation in the reading assignment above pertaining to $(G, *)$. Save your definition for reference.

3.7   A BRIEF LOOK AT ISOMORPHISM

Look at Table 3.9 for the operational system $(A, @)$, where A is the set $A = \{a, b, c, d\}$:

TABLE 3.9

@	a	b	c	d
a	a	b	c	d
b	b	c	d	a
c	c	d	a	b
d	d	a	b	c

PROBLEM 1   a. Is $(A, @)$ a group? (Check only three examples of associativity to answer the question or check all sixty-four examples via a computer program.)

b. List the identity element, $e =$ _____ .

c. Next to each element, list its inverse, if any exists.

$a' =$ _____

$b' =$ _____

$c' =$ _____

$d' =$ _____

PROBLEM 2   **a.** Try to find an operational system studied previously which resembles $(A, @)$. (Look at the patterns in the table)
     b. Try rewriting the table for $(A, @)$, relabeling elements of $(A, @)$ as follows:

a relabeled as 0

b relabeled as 1

c relabeled as 2

d relabeled as 3

     c. What is the relabeled operational system?

PROBLEM 3   **a.** Since the identity of $(A, @)$ is the element a, we would expect the identity of the relabeled system to be 0. Is it?
     b. The element b of $(A, @)$ has been relabeled as 1. We might expect b' to be replaced by $1'$ in the new table. Is it?

PROBLEM 4   **a.** Under the relabeling indicated in Problem 2, you will note that $c \leftrightarrow 2$ and $d \leftrightarrow 3$. Does $c @ d \leftrightarrow 2 + 3$?
     b. Check several other examples to see if this works.
     c. Do you think that this will always work for $(A, @)$ and $(Z_4, +)$?

You will probably recognize the two groups given in Tables 3.10 and 3.11 as the motions of the equilateral triangle under composition and mod 6 addition.

TABLE 3.10

			(M, o)			
o	R_3	R_1	R_2	F_T	F_R	F_L
R_3	R_3	R_1	R_2	F_T	F_R	F_L
R_1	R_1	R_2	R_3	F_R	F_L	F_T
R_2	R_2	R_3	R_1	F_L	F_T	F_R
F_T	F_T	F_L	F_R	R_3	R_2	R_1
F_R	F_R	F_T	F_L	R_1	R_3	R_2
F_L	F_L	F_R	F_T	R_2	R_1	R_3

It is reasonable to ask if the elements of (M, o) can be relabeled so that the operation table will be the same as the $(Z_6, +)$

TABLE 3.11

$(Z_6, +)$

+	0	1	2	3	4	5
0	0	1	2	3	4	5
1	1	2	3	4	5	0
2	2	3	4	5	0	1
3	3	4	5	0	1	2
4	4	5	0	1	2	3
5	5	0	1	2	3	4

operation table. Of all the possible relabeling schemes, those that match each element of M with *exactly one* element of Z_6 are of interest. In these cases the elements of the sets are in *one-to-one correspondence*. There are 720 possible one-to-one correspondences between the elements of these two sets. One of these possible schemes for relabeling follows:

$$R_3 \leftrightarrow 0 \quad F_T \leftrightarrow 3$$
$$R_1 \leftrightarrow 1 \quad F_R \leftrightarrow 4$$
$$R_2 \leftrightarrow 2 \quad F_L \leftrightarrow 5$$

This isn't a completely arbitrary choice since if the relabeled operation table is to be the same as the $(Z_6, +)$ operation table, certainly the identity elements of the two groups must be matched to each other. In the above scheme the identities are matched.

PROBLEM 5 F_T is matched with 3; do their respective inverses match?

PROBLEM 6 $R_1 \leftrightarrow 1$ and $R_2 \leftrightarrow 2$; does $R_1 \circ R_2 \leftrightarrow 1 + 2$?

Your answer to Problem 6 should have been no. This may be somewhat surprising to you, but it really shouldn't be. Remember this is only one of 720 different possible one-to-one correspondences.

PROBLEM 7   Determine whether each of the following one-to-one correspondences

98

will relabel the (M, o) operation table to be the same as the operation table for $(Z_6, +)$.

	a.	b.	c.
	$R_3 \leftrightarrow 0$	$R_3 \leftrightarrow 4$	$R_3 \leftrightarrow 0$
	$R_1 \leftrightarrow 2$	$R_1 \leftrightarrow 2$	$R_1 \leftrightarrow 4$
	$R_2 \leftrightarrow 1$	$R_2 \leftrightarrow 0$	$R_2 \leftrightarrow 5$
	$F_T \leftrightarrow 4$	$F_T \leftrightarrow 3$	$F_T \leftrightarrow 1$
	$F_R \leftrightarrow 3$	$F_R \leftrightarrow 5$	$F_R \leftrightarrow 2$
	$F_L \leftrightarrow 5$	$F_L \leftrightarrow 1$	$F_L \leftrightarrow 3$

PROBLEM 8   To actually prove that there is no relabeling of (M, o) that will make the operation table the same as $(Z_6, +)$ you would either have to check the remaining 716 cases, or find a basic difference between the structure of the two systems. Try to find such a difference to show that there is no relabeling that works.

Because $(A, @)$ can be relabeled in such a way that the operation table is the same as $(Z_4, +)$, the two groups are called *isomorphic* and the one-to-one correspondence is called an *isomorphism*. As the word implies, an isomorphism between two groups simply means that the groups have exactly the same structure. (M, o) and $(Z_6, +)$ are not isomorphic. To make the idea of isomorphism precise, the following definition is made:

DEFINITION: Two finite groups, $(G, *_1)$ and $(\hat{G}, *_2)$ are said to be isomorphic if and only if the two conditions listed below both hold:

1. A one-to-one correspondence can be established between the elements of G and $\hat{G}$:

$$G \quad \hat{G}$$
$$a \leftrightarrow \hat{a}$$
$$b \leftrightarrow \hat{b}$$
$$c \leftrightarrow \hat{c}$$
$$\vdots$$

2. For all $a, b, c \in G$ and $\hat{a}, \hat{b}, \hat{c} \in \hat{G}$, if $a *_1 b = c$, then $\hat{a} *_2 \hat{b} = \hat{c}$.

PROBLEM 9   Establish a one-to-one correspondence between $(Z_2, +)$ and $(P, *)$ and show that it is an isomorphism by the above definition.

TABLE 3.12

$(Z_2, +)$

+	0	1
0	0	1
1	1	0

TABLE 3.13

$(P, *)$

*	a	b
a	b	a
b	a	b

You now have a definition for isomorphism. This concept that lays bare the same form, in different systems, is a powerful idea. For instance, the isomorphism between adding numbers and moves on the number line enables a teacher to use the number line as an excellent teaching device. It is more than just a mere analogy. What happens in one system also happens in the other system.

HOMEWORK EXERCISES

1. Refer to your operation table for the group of motions of a rectangle, (X, o). (Problem 17, Section 3.5).
   a. Does the table for (X, o) show the same pattern as the table for $(A, @)$?
   b. To answer this, you might try listing each element of (X, o) next to its inverse. What do you observe?
   c. Are these two groups isomorphic?

2. List the tables for $(Z_8, +)$ and (S, o), where (S, o) is the group of motions of the square under the operation composition.
   a. What differences do you notice between these two groups?
   b. Are $(Z_8, +)$ and (S, o) isomorphic? Why or why not?

3. Are the following isomorphic?
   If so, give a possible 1-to-1 correspondence.
   If not, tell why.
   a. $(Z_3, +)$ and $(Z_3, \cdot)$
   b. $(Z_4, +)$ and $(Z_5 - \{0\}, \cdot)$
   c. (X, o) and $(Z_4, +)$ (X is the same set as in Exercise 1.)
   d. (Rigid motions of regular hexagon, o) and $(Z_{12}, +)$

4. Is $(Z, +)$ isomorphic to $(S, +)$ where

$$S = \left\{ \cdots \frac{1}{-3}, \frac{1}{-2}, \frac{1}{-1}, 0, \frac{1}{1}, \frac{1}{2}, \frac{1}{3} \cdots \right\}$$

If so, give a 1-to-1 correspondence.
If not, tell why.

★3.8   MORE EXAMPLES OF FINITE GROUPS

Suppose that you have a set B of four objects:

a blue triangle,
a blue circle,
a red triangle,
a red circle.

If available, attribute blocks will be helpful. If they are not available, represent these four objects in some manner. You could define a "color-changing function" which will change each block in B to the block of the same shape but opposite color.

PROBLEM 1   If you start with the blocks lined up as shown in Figure 3.14, what change will this function C make? Draw the result.

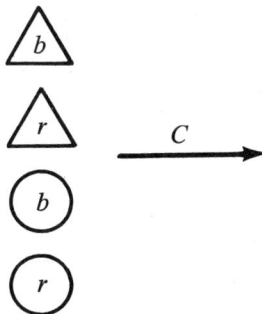

FIGURE 3.14

You could also define S, a "color-changing function," that will change each block in *B* to the block of the same color but opposite shape.

PROBLEM 2   What change will the function *S* make in the original display of blocks? Draw the result.

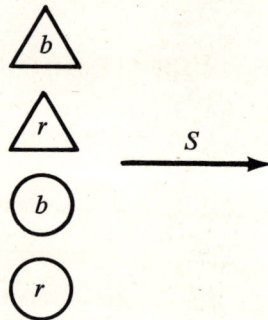

FIGURE 3.15

Let the letter *I* represent the no-change function (which leaves all blocks in their original position). You now have three different functions defined on *B* so far.

PROBLEM 3   a. See what change *S* o *C* makes on the original display of blocks where o means *followed by*.
b. How about *C* o *S*?

This should give a total of four different functions which are defined on *B*. Call this set of functions *F*.

PROBLEM 4   a. Make an operation table for (*F*, o).
b. Prove or disprove: (*F*, o) is a commutative group. (Associativity can be checked via computer program.)
c. Examine the table for (*F*, o) and find a familiar group to which (*F*, o) is isomorphic.

Problems 1 through 4 adapted from Zolton P. Dienes, *Building Up Mathematics* (New York: Hutchinson Educational Ltd., 1969), pp. 144–145.

For a second example, let U be the set $U = \{1, 2\}$, and let $P(U)$ represent the set consisting of all subsets of U.

Recall $P(U) = \{\{1, 2\}, \{1\}, \{2\}, \{\}\}$.

Define the operation $\wedge$ as follows:

$A \wedge B = (A \cap \bar{B}) \cup (\bar{A} \cap B)$ where A, B are elements of $P(U)$.

For example, $\{1, 2\} \wedge \{1\} = (\{1, 2\} \cap \{2\}) \cup (\{\} \cap \{1\})$
$= \{2\} \cup \{\} = \{2\}$.

PROBLEM 5   **a.** Show by means of a table that $(P(U), \cap)$ is a commutative group.

b. Is $(P(U), \cap)$ isomorphic with (F, o) from the previous example?

HOMEWORK EXERCISES

1. **a.** Write the operation table for a group with two elements which has the same structure as $(Z_2, +)$. Use the set $S = \{\bigcirc, \square\}$.

   b. Now try to find a group with two elements which is *not* isomorphic to $(Z_2, +)$. What do you conclude?

2. **a.** Complete the operation table for the set $\{1, 2, 3, 4\}$ under the operation multiplication mod 5.

TABLE 3.14

·	1	2	4	3
1				
2				
4				
3				

Complete Table 3.15 for the set $\{1, 3, 9, 7\}$ under the operation multiplication mod 10.

TABLE 3.15

·	1	3	9	7
1				
3				
9				
7				

  b. Show that each operational system in part **a** is a commutative group.
  c. Are these two groups isomorphic?
  d. Are they isomorphic to any familiar groups?
  e. If so, which groups?

3.9 EXPLORING OTHER PROPERTIES OF GROUPS

PROBLEM 1 List several examples of groups and refer to them in answering the following questions.
  a. How many identity elements are in each group?
  b. What is the inverse of the identity element in each group?
  c. Does any element have more than one inverse in any group?
  d. Pick an element, a, out of each group. Find a'. Is it an element of the group? is $(a')'$ an element of the group? What is $(a')'$ equal to?

PROBLEM 2 Look at each of your answers in Problem 1. Generalize a little bit and construct four conjectures that you think will be true in *any* group. Write each of these conjectures.

PROBLEM 3 Complete the conjecture:
In a group G, if $a * c = b * c$, then _____.

PROBLEM 4 Complete the conjecture:
In a group G, if $a * x = b$, then $x =$ _____ .

PROBLEM 5 Test your conjecture in Problem 4 by referring to the table for (M, o), the group of motions of the equilateral triangle under the operation of composition.
  a. First, write the conjecture using the notation of (M, o).
  In the group (M, o) if $a o x = b$, then $x =$ _____ .
  b. Now let $a = F_T, b = R_2$ in the sentence $a o x = b$. Find x.
  c. Try one more. Let $a = R_1$ and $b = F_L$ and find x. If your conjecture is not correct, revise your conjecture and try again.

PROBLEM 6 In this problem you are seeking an expression for the inverse of a composition, i.e., an expression for $(a * b)'$.

To give you an idea, refer to the table for (M, o), the group of motions of the equilateral triangle under the operation of composition.

Let $a = R_2$, $b = F_R$.

a. First compute $(R_2 \text{ o } F_R)'$. Now find $(R_2' \text{ o } F_R')$ and $(F_R' \text{ o } R_2')$. Which of these is the same group element as $(R_2 \text{ o } F_R)'$?

b. Repeat the process above to find an expression for $(F_L \text{ o } F_T)'$.

c. Now complete the conjecture: In a group G, $(a * b)' = \underline{\hspace{1.5cm}}$.

3.10 SOME THEOREMS ABOUT GROUPS

Your explorations have led you to some conjectures. Now these need to be proven in an arbitrary group $(G, *)$. Once a conjecture has been proven, it is called a *theorem*.

The most basic of these conjectures is the *right cancellation law:*

In a group G, if $a * c = b * c$ then $a = b$, for all $a, b, c \in G$.

PROBLEM 1   The statements in the proof of the right cancellation law follow. Your job is to fill in the reasons.

Prove: In a group G, if $a * c = b * c$ then $a = b$, for all $a, b, c \in G$.

Proof:	*Statements*	*Reasons*
1.	$a, b, c \in G$	1.
2.	$a * c = b * c$	2.
3.	There exists $c' \in G$	3.
4.	$(a * c) * c' = (b * c) * c'$	4.
5.	$a * (c * c') = b * (c * c')$	5.
6.	$a * e = b * e$	6.
7.	$a = b$	7.

Note: In the proof of the right cancellation law you used *all* of the properties in the definition of a group.

105

HOMEWORK EXERCISES

1. Prove the *left cancellation law*, i.e.: In a group G, if $c * a = c * b$ then $a = b$, for all $a, b, c \in G$.

2. Consider the statement, in a group G, if $a * c = c * b$ then $a = b$, for all $a, b, c \in G$. Is this true? If so, prove it. If not, explain why.

PROBLEM 2   Prove that a group has exactly one identity element, i.e., the identity element is unique. Hint: Suppose that e and e_1 are both identity elements for G. Then for all $a \in G$, $a * e = a$ and $a * e_1 = a$. Now prove that $e = e_1$.

PROBLEM 3   Each element in a group has exactly one inverse. For the proof, assume that an element a has two inverses, a' and a_1'. Prove that $a' = a_1'$.

PROBLEM 4   The inverse of the identity is the identity. Prove that $e' = e$. Hint: Use e as the identity and also as an element whose inverse is e'.

PROBLEM 5   The inverse of the inverse of an element is the element. In G, $(a')' = a$, for all $a \in G$. Prove it. Remember that $(a')'$ is the inverse of a'.

PROBLEM 6   The inverse of a composition is equal to the composition of the inverses in reverse order. In G, $(a * b)' = b' * a'$, for all $a, b \in G$. Prove this, using the fact that $(a * b)' * (a * b) = e$.

PROBLEM 7   Prove that for all $a, b \in G$, the equation $x * a = b$ has a unique solution for x in G. Hint: This proof has two parts:
   a. First you must find a value for x and check it by substituting it into the equation. This proves the existence of a value for x.
   b. The second part requires you to show that the solution is unique.

Proofs are an important part of mathematics. This is the deductive side of the subject. For the proofs in this section you started with the definition of a group. From the properties stated in this definition, you were able to prove deductively other properties of a group. This is just a very small sample of proof. Some courses in advanced mathematics begin with a set of propositions and the material for the entire course is proved from these propositions. Mathematics serves as an excellent model for this approach.

HOMEWORK EXERCISES

3. This problem will show the connection between the characteristics of a general group and those of familiar groups with operations of addition and multiplication.

    For each of the preceding theorems (Problems 1 to 7), restate the theorem in two ways:
    a. for an arbitrary additive group $(G, +)$
    b. for an arbitrary multiplicative group $(G, \cdot)$

4. For all $a, b \in G$, if $a * y = b$, is $y = b * a'$? (This is *not* just like Problem 7.) Why or why not? If not, find a replacement for y which makes the equation $a * y = b$ true.

5. a. Complete and prove: For all $a, b \in G$, if $a' = b$, then $b' = $ _____.
    b. Prove: If $(a * b)^2 = a^2 * b^2$ for all $a, b \in G$, then G is commutative. Remember the definition of "squared." $x^2 = x * x$.
    ★c. Prove: If $a = a'$ for all $a \in G$, then G is commutative.
    ★d. Is the converse true? i.e., If G is commutative, does $a = a'$ for all $a \in G$?

3.11 A FEW INFINITE GROUPS AND A SUMMARY

You have now looked at a number of groups with different operations in finite sets, but no infinite sets were included. Review the properties that hold in whole numbers, integers, and rational numbers under addition and multiplication.

PROBLEM 1   Answer the following questions for each of the sets W, Z, Q.
    a. Is it a group under addition? under multiplication?
    b. Can you prove your decision or only make a conjecture?
    c. Is $Q - \{0\}$ a group under addition? under multiplication?

HOMEWORK EXERCISES

1. a. Based on your exploration to date, complete Table 3.16 by writing Yes or No.

TABLE 3.16

System	Rectangle (X, o)	Equilateral Triangle (M, o)	Square (S, o)	$(W, +)$	$(W, \cdot)$	$(Z, +)$	$(Z, \cdot)$
Group?							

System	$(Q, +)$	$(Q, \cdot)$	$(Q - \{0\}, \cdot)$	$(Z_n{}' +)$	$(Z_n - \{0\}, \cdot)$ n prime	$(Z_n - \{0\}, \cdot)$ n not prime
Group?						

b. Which of the groups in Table 3.16 are commutative?

It would be reasonable to ask why one bothers to check whether a set under a given operation is a group. One reason is to use the theorems you proved previously without having to prove them again. Since they were proven for groups in general, the theorems may now be used in any particular group.

DEFINITION: A group G is a set of elements and a binary operation $*$ with the following properties:

1. *Closure:*   For all a, b in G, $a * b$ is in G.
2. *Associativity:*   For all a, b, c in G, $a * (b * c) = (a * b) * c$.
3. *Identity Element:*   There is an element e in G such that $a * e = e * a = a$, for all a in G.
4. *Inverse Elements:*   For each a in G, there exists an a' in G such that $a * a' = a' * a = e$.

Theorems that apply to all groups:

1. *Cancellation Laws:*   For all a, b, c in G, $a * b = a * c$ implies $b = c$ and $a * b = c * b$ implies $a = c$.
2. *Unique Identity:*   A group has exactly one identity element.
3. *Unique Inverses:*   Each element of a group has exactly one inverse.

4. *Inverse of the Identity:* In *G*, *e'* = *e*.
5. *Inverse of the Inverse:* For all *a* in *G*, (*a'*)' = *a*.
6. *Inverse of* (*a* * *b*): For all *a*, *b*, in *G*, (*a* * *b*)' = *b'* * *a'*.
7. *Unique Solutions for Equations:* For all *a*, *b* in *G*, *a* * *x* = *b* has the unique solution *x* = *a'* * *b* and *y* * *a* = *b* has the unique solution *y* = *b* * *a'*.

★3.12   MORE ON ISOMORPHISM

You have had a brief introduction to isomorphism in Section 3.7. This section will use the concept of isomorphism to continue the focus on the processes of mathematics. Here you will develop a tool to find an isomorphism between two systems. Then you will explore patterns, formulate conjectures, and prove statements. These processes are more important than the specific results.

PROBLEM 1   Construct an operation table for (Z_4, +) and for (Z_5 − $\{0\}$, ·). Z_5 − $\{0\}$ refers to the set of elements $\{1, 2, 3, 4\}$, and, in this case, · refers to multiplication mod 5.

By just comparing tables, it's hard to determine whether these two groups (Z_4, +) and (Z_5 − $\{0\}$, ·) are isomorphic. Examine the tables more closely. First look at (Z_4, +). You'll note that each element in Z_4 can be obtained by adding the element 1 to itself an appropriate number of times.

$$1 = 1$$
$$2 = 1 + 1$$
$$3 = 1 + 1 + 1$$
$$0 = 1 + 1 + 1 + 1$$

The element 1 is called the *generator* of this group. (Z_4, +) is called a *cyclic* group because a single element can be used to generate each element in the group. Now $g = 1, g^2 = 2, g^3 = 3, g^4 = 0$. The exponent is generally used to indicate repeated use of the generator with the operation, even though, in this case, the operation is addition. Since there is never more than one operation in a group, this will cause no difficulty.

109

PROBLEM 2   Rewrite Table 3.17 for $(Z_4, +)$ placing the generator and its powers in order.

TABLE 3.17

. +	$g = 1$	$g^2 = 2$	$g^3 = 3$	$g^4 = 0$
$g = 1$				
$g^2 = 2$				
$g^3 = 3$			2	
$g^4 = 0$	1			

PROBLEM 3   Does $(Z_5 - \{0\}, \cdot)$ have a generator? i.e., is there an element g in $Z_5 - \{0\}$ whose powers $g^1, g^2, g^3, \ldots$, yield all the elements of $Z_5 - \{0\}$? Look carefully, it may not be the element 1.

Now use generators to establish an isomorphism between $(Z_4, +)$ and $(Z_5 - \{0\}, \cdot)$.

PROBLEM 4   To see this more clearly, rearrange your table for $(Z_5 - \{0\}, \cdot)$ in terms of powers of the element which you found to be the generator of the group. Let g represent the generator.

TABLE 3.18

.	g	g^2	g^3	g^4
g				
g^2				
g^3				
g^4				

PROBLEM 5   Now establish an isomorphism between $(Z_4, +)$ and $(Z_5 - \{0\}, \cdot)$. Compare the tables in Problems 2 and 4. Remember there are two parts to the definition of isomorphism—first establish a one-to-one correspondence and then show that results correspond when operating on corresponding elements in the two groups.

PROBLEM 6   a.   Construct tables for $(Z_6, +)$ and for $(Z_7 - \{0\}, \cdot)$.
   b.   Show that each of these two groups is cyclic, i.e., for each group find a single element which generates the group.

c.   Then show that the two groups are isomorphic.

★PROBLEM 7   There is more than one isomorphism between $(Z_6, +)$ and $(Z_7 - \{0\}, \cdot)$. Establish a different isomorphism by using a different element as the generator of $(Z_7 - \{0\}, \cdot)$.

PROBLEM 8   a.   Could you do the same thing as in Problem 6 for $(Z_5, +)$ and $(Z_6 - \{0\}, \cdot)$?
b.   How about for $(Z_2, +)$ and $(Z_3 - \{0\}, \cdot)$?
c.   Now compare the three cases in which you could establish the isomorphism. Can you make any conjectures about an isomorphism between $(Z_{n-1}, +)$ and $(Z_n - \{0\}, \cdot)$? If not, ask your instructor for more hints.

★PROBLEM 9   Let T represent the set of elements of $(Z_{10}, \cdot)$ which can be represented as powers of the element 3. For example, 7 is an element of T since $3^3 = 3 \cdot 3 \cdot 3 = 7$, but 2 is not an element of T. Find the other elements of the set T. Make a table for the set T under the operation multiplication mod 10 and determine whether $(T, \cdot)$ is a group.

★PROBLEM 10   Use what you know about generators to show that $(T, \cdot)$ and $(Z_5 - \{0\}, \cdot)$ are isomoprhic. To what other familiar group are they isomorphic?

Now investigate the possibility of establishing isomorphisms between a few infinite groups. The definition will be the same as the one for finite groups in Section 3.7. You have already shown that $(Z, +)$ is a group. Now let's consider the set D of integers which are multiples of 3 under the operation of ordinary addition. You could write D as follows:
$D = \{ x \mid x = 3 \cdot z, z \in Z \}$. This is read "$D$ equals the set of all x such that x equals 3 times z, z is an element of Z."

PROBLEM 11   a.   List a dozen elements of D.
b.   Is $(D, +)$ a group? Why or why not?

When you established an isomorphism between finite groups, you first showed that one system was merely a relabeling of the other. To do this, you showed that you could match each element in one group with an element in the other, i.e., establish a one-to-one correspondence between the elements of the two groups.

To show that two finite groups were isomorphic, you then showed that the operation tables for the two groups were identical except for relabeling. In the infinite case, you can't compare tables; but you can show that the operations behave "similarly" by applying the second part of the definition of isomorphic.

PROBLEM 12   Is there a natural way to match elements of $(Z, +)$ with elements of $(D, +)$? To indicate the way to match elements, list some elements of Z on one line and list the corresponding elements of D below them.

PROBLEM 13   Refer to the two lines of corresponding elements which you constructed for Z and D. Choose two of the elements, a and b, from Z which you have listed and add them. Let $s = a + b$ represent the sum which you obtain, $s =$ _____. Now to check whether the operations behave similarly, add the two corresponding elements, a_1 and b_1, of D and see whether this sum is the element below s. Try this procedure for several other pairs of elements in Z and D.

PROBLEM 14   If $m + n = k$ in $(Z, +)$, would you expect that $3 \cdot m + 3 \cdot n = 3 \cdot k$? Why? Explain why this is a restatement of the procedure you carried out in Problem 13.

★PROBLEM 15   Now choose a nonzero integer as a value for n and show that $(Z, +)$ and $(nZ, +)$ are isomorphic. The set nZ is defined as follows:

$$nZ = \{ x \mid x = n \cdot z, z \in Z \}.$$

First list some elements of nZ.

★PROBLEM 16   Is $(Z, +)$ isomorphic to $(nZ, +)$ for every nonzero integer n? Why or why not?

HOMEWORK EXERCISES

1. Could a group with four elements be isomorphic to both $(Z_4, +)$ and to the group of motions of the rectangle under the operation composition simultaneously? Why or why not?

2. Show that $(Z, +)$ is isomorphic to $(B, +)$ where $B = \left\{ y \mid y = \dfrac{k}{7}, k \in Z \right\}$.

3. In $(Z_{21}, \cdot)$, find the set of elements K which can be generated from the

element 5. Show that $(K, \cdot)$ is a commutative group, where $\cdot$ represents multiplication mod 21. Find a familiar group to which $(K, \cdot)$ is isomorphic.

★3.13 SUBGROUPS

DEFINITION: Let $(G, *)$ be a group. Then $(H, *)$ is a *subgroup* of $(G, *)$ if and only if H is a nonempty subset of G and $(H, *)$ is a group. Study the examples below to more clearly understand the definition.

Note: Some subsets form subgroups and some do not.

For example, look at $(Z_2, +)$.

TABLE 3.19

+	0	1
0	0	1
1	1	0

The nonempty subsets of Z_2 are $\{0\}$, $\{1\}$, $\{0, 1\}$. Hence the possibilities for subgroups are shown in the following tables:

TABLE 3.20

+	0
0	0

TABLE 3.21

+	1
1	0

TABLE 3.22

+	0	1
0	0	1
1	1	0

Because the systems are groups, $(\{0\}, +)$ & $(\{0, 1\}, +)$ are subgroups of $(Z_2, +)$.

PROBLEM 1   Why isn't the system $(\{1\}, +)$ a subgroup of $(Z_2, +)$?

PROBLEM 2   Now, consider $(Z_4, +)$.

TABLE 3.23

+	0	1	2	3
0	0	1	2	3
1	1	2	3	0
2	2	3	0	1
3	3	0	1	2

List all nonempty subsets of Z_4.

Check to see which subsets are groups under mod 4 addition; these are the subgroups of $(Z_4, +)$. (It is not necessary to make operation tables.)

PROBLEM 3   a.   Is $(\{^-1, 0, 1\}, +)$ a subgroup of $(Z, +)$? Why or why not?

b.   Let $(H, +)$ be a subgroup of $(Z, +)$ such that $3 \in H$. Find a dozen other elements of H. Is H finite or infinite?

PROBLEM 4   a.   How many possibilities are there for subgroups of a finite group with n elements? (Actually, the number of subgroups turns out to be far less than this number.)

b.   Is it possible in a finite group to have a subgroup with an infinite number of elements?

c.   Is it possible to have a subgroup of an infinite group with an infinite number of elements?

d.   Is it possible to have a finite subgroup of an infinite group?

Problems 5 and 6 will give you some clues as to how to form subgroups.

PROBLEM 5   If e is the identity element of a group $(G, *)$, is e necessarily the identity element of *every* subgroup? Justify.

PROBLEM 6   In any group $(G, *)$, are $(\{e\}, *)$ and $(G, *)$ necessarily subgroups of $(G, *)$?

DEFINITION: The *order* of a group (or subgroup) $(G, *)$ is the number of elements in G. Order can be thought of as the cardinality of G if this is more familiar to you.

One of the fundamental results in group theory is the relationship between the order of a finite group and the order of each

114

of its subgroups. The next few problems are intended to lead to the discovery of this relationship.

PROBLEM 7   Find all the subgroups of $(Z_6, +)$ and then find their orders.

PROBLEM 8   Find all the subgroups of each of the following: $(Z_2, +)$, $(Z_3, +)$, ... $(Z_{10}, +)$. Use Problems 1, 2, and 7 to give you a start. Again, don't make tables unless you need them. Record your results by completing Table 3.24.

TABLE 3.24

Group	Subgroups	Order of Group	Order(s) of Subgroup
$(Z_2, +)$	$(\{0\}, +), (\{0, 1\}, +)$	2	1, 2
$(Z_3, +)$			
$(Z_4, +)$			
$(Z_5, +)$	$(\{0\}, +),$ $(\{0, 1, 2, 3, 4\}, +)$		1, 2, 4
$(Z_6, +)$			
$(Z_7, +)$			
$(Z_8, +)$			
$(Z_9, +)$			
$(Z_{10}, +)$			

PROBLEM 9   Look closely at the last two columns of the table in Problem 8 and then complete the following conjecture relating the order of a group to the order of each of its subgroups:

If $(H, *)$ is a subgroup of $(G, *)$, then _____.

Check this with your instructor. What you have probably stated (if you didn't goof!) is called Lagrange's theorem.

PROBLEM 10   Find all the subgroups of (M, o) (Problem 8, Section 3.5).

HOMEWORK EXERCISES

1. Find at least 2 infinite subgroups of $(Q, +)$.

2. Give an example to show that a noncommutative group may have a commutative subgroup.

3. If $(K, *)$ is a subgroup of $(H, *)$ and $(H, *)$ is a subgroup of $(G, *)$, is $(K, *)$ necessarily a subgroup of $(G, *)$?

4. Find all the subgroups of $(Z_{30}, +)$.

★5. If $(H, *)$ and $(K, *)$ are subgroups of $(G, *)$, is $(H \cap K, *)$ necessarily a subgroup of G? Justify your answer.

SUMMARY

We have explored the important unifying concept of binary operation. Binary operations were used to structure different sets of numbers, sets of points, sets of motions of geometric figures, and even sets of sets. Some of these binary operations were different from the usual operations of addition, subtraction, multiplication, and division. The most important structure examined in these many sets was the structure of a group. Groups are a structure imposed on a set by one operation. Many different kinds of sets and a variety of different operations all yield the group structure. A new and powerful concept of isomorphism was introduced to show when structures that may look radically different are actually of the same basic form. Once again the cycle of mathematical processes was used in developing these concepts: pattern hunting, rule formation, and pattern testing. Finally, statements about groups were examined to begin to develop a deductive system of theorems.

Sets Structured by Relations

4

4.1 RELATIONS AND THEIR GRAPHS

PROBLEM 1   Figure 4.1 shows some of the relationships that involved Henry VIII.

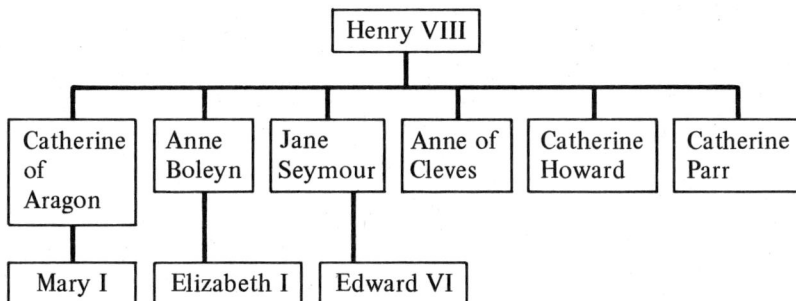

```
                          ┌──────────┐
                          │ Henry VIII│
                          └──────────┘
  ┌──────────┬──────────┬──────────┬──────────┬──────────┬──────────┐
┌──────────┐┌────────┐┌────────┐┌────────┐┌──────────┐┌──────────┐
│Catherine ││Anne    ││Jane    ││Anne of ││Catherine ││Catherine │
│of        ││Boleyn  ││Seymour ││Cleves  ││Howard    ││Parr      │
│Aragon    ││        ││        ││        ││          ││          │
└──────────┘└────────┘└────────┘└────────┘└──────────┘└──────────┘
┌──────────┐┌──────────┐┌──────────┐
│ Mary I   ││Elizabeth I││Edward VI │
└──────────┘└──────────┘└──────────┘
```

FIGURE 4.1

How are Henry VIII and Catherine of Aragon related?
How are Jane Seymour and Henry VIII related?
How are Elizabeth I and Anne Boleyn related?
How are Henry VIII and Mary I related?
Name a few other relationships.

PROBLEM 2   Suppose you look at the following set of ordered pairs: $H = \{$ (Henry

VIII, Anne B.), (Henry VIII, Catherine of Aragon), (Henry VIII, Jane Seymour), (Henry VIII, Anne C.), (Henry VIII, Catherine Parr), (Henry VIII, Catherine Howard) }.

What criterion, or rule, describes the pairs that belong to *H*? Would (Jane Seymour, Henry VIII) be an appropriate ordered pair to be included in *H*? Why, or why not? Be sure when you explore this last question that you refer to the criterion that you decided upon. The ordered pair (Jane Seymour, Henry VIII) might be an appropriate ordered pair under one criterion and inappropriate under another criterion.

Suppose in Problem 2 you chose the criterion, *was the husband of,* where the two members of the ordered pair can be used to complete the sentence: Henry VIII was the husband of Anne B. In this case, the ordered pair (Jane Seymour, Henry VIII) does not belong to the set of ordered pairs. To say that Jane Seymour was the husband of Henry VIII is simply nonsense.

Suppose in Problem 2 above you chose the criterion, *was the spouse of.* Now the ordered pair (Jane Seymour, Henry VIII) does belong to the set of ordered pairs.

PROBLEM 3   Let *S* be the set of ordered pairs for the relation, *was the spouse of.* List all the ordered pairs that would belong to the relation *S.* (Abbreviate to save writing.)

PROBLEM 4   Use the same set of people that are related to Henry VIII. Now define the relation *D* with the criterion, *was the daughter of.* List all the ordered pairs of *D.*

HOMEWORK EXERCISES

1. Give the set of ordered pairs for the criterion defined on the given set:

   P = { Henry VIII, Catherine of Aragon, Mary I, Anne B., Elizabeth I, Jane Seymour, Edward VI, Anne of Cleves, Catherine H., Catherine Parr }

   a.   *. . . was the wife of . . .*
   b.   *. . . was the father of . . .*
   c.   *. . . was the half brother of . . .*
   d.   *. . . was the half sister of . . .*

PROBLEM 5   Now look at the set:

K = { Jane Smith, Stan West, Scott Moyer, Nancy Johnson, Bill Webb, Al Anderson, William Blake } .

In Figure 4.2, label the remaining points so that each element of K is represented by a single point.

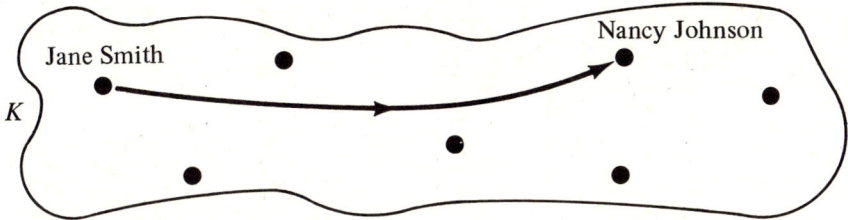

FIGURE 4.2

Now consider a certain relationship among the members of set K. Whenever the first initial of a person's first name is the same as the first initial of a person's last name, draw an arrow from the first person to the second person (as in Figure 4.2). Such a pictorial representation of a relation is called a *directed graph*. Should your directed graph have an arrow from Nancy Johnson to Jane Smith? Why or why not? What, if anything, should be done about Al Anderson?

PROBLEM 6   The directed graph, Figure 4.3, was drawn to represent a first name–last name relation for some set of persons. Try to choose initials so that no extra arrows will be needed. Explain why or why not the graph is possible.

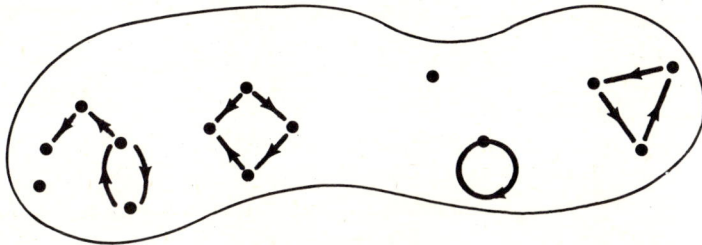

FIGURE 4.3

119

HOMEWORK EXERCISES

2. Decide whether the directed graph in Figure 4.4 can be a correct graph for the first name–last name relation defined on a set of persons. If it can be, write appropriate initials for each point. If it cannot be, draw in additional arrows to make it a correct graph.

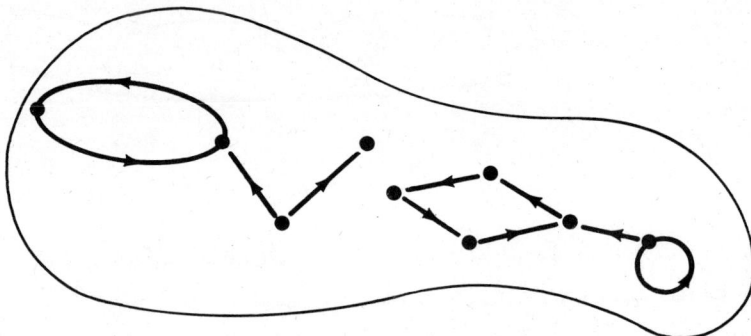

FIGURE 4.4

3. The directed graph in Figure 4.5 represents the relation, *is a sister of,* on a set of boys and girls. Determine which points are girls and which points are boys.

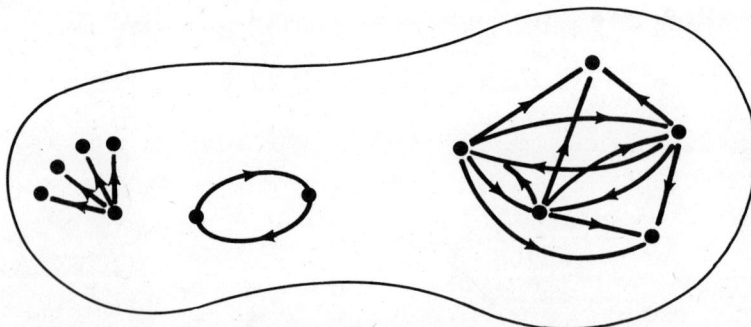

FIGURE 4.5

PROBLEM 7   Now look at set $P = \{$Brenda Jacobs, Dave Smith, Jim Johnson, Diane Adams, John Snyder $\}$ and consider the relation, *has the same first name initial as,* defined on set *P*. Draw a directed graph to represent this relation on set *P*.

The directed graph that you have drawn represents a relation on the set *P*. Whenever an arrow goes from *a* to *b*, we say that the ordered pair (*a, b*) is an element of the relation. Hence, a relation is a set whose elements are ordered in pairs.

DEFINITION: Given a set *S*, a *relation* is any set of ordered pairs of the elements of set *S*.

PROBLEM 8   Now consider a relation involving integers. Let $S = \{1, 2, 3, 4, 5\}$. Then, one relation on set *S* is $R = \{(1, 2), (1, 3)\}$ since *R* is a set consisting of ordered pairs of the elements of sets. Another relation on *S* is found by using the criterion, *is less than.* Draw a directed graph to represent the relation, *is less than,* on *S*.

Denote the relation in Problem 8 by $<$. You know that $(1, 3) \in <$ since 1 is less than 3. However this is usually written $1 < 3$. In fact, this notation can be extended for all relations. If *R* is any relation, and (*a, b*) $\in R$, we can write *a R b*.

PROBLEM 9   Using the relation $<$ on the set of integers, *Z*, write several ordered pairs that are elements of $<$. Can any element of *Z* be related to itself? Suppose $a < b$. Is it true that $b < a$? Suppose $a < b$ and $b < c$. Is it true that $a < c$?

Another relation, *is a multiple of,* can be defined on the set of integers.

DEFINITION: For p, $q \in Z$, p is a multiple of q if and only if there exists an integer m such that $p = q \cdot m$.

PROBLEM 10   Using the relation, *is a multiple of,* on the set of integers, write several ordered pairs of integers that are elements of the relation. Are the elements of Z related to themselves? If a is a multiple of b, is b a multiple of a? If a is a multiple of b and b is a multiple of c, would a be a multiple of c?

PROBLEM 11   Now consider a set consisting of boys and girls, and the relation, *is a sister of.* If a directed graph is made for this relation, could the graph have any arrows that begin and end at the same point? If an arrow goes from a to b, must there also be an arrow from b to a? Why or why not? Suppose an arrow goes from a to b, and one goes from b to c. Must there also be an arrow from a to c?

PROBLEM 12   Now consider a set which consists only of girls, and the relation, *is a sister of.* Make up your own directed graph to represent the relation. If an arrow goes from a to b, must one go from b to a? If an arrow goes from a to b, and one goes from b to c, must there also be an arrow from a to c?

PROBLEM 13   Figure 4.6 is the graph of the relation, *is a sister of,* on a set of girls.

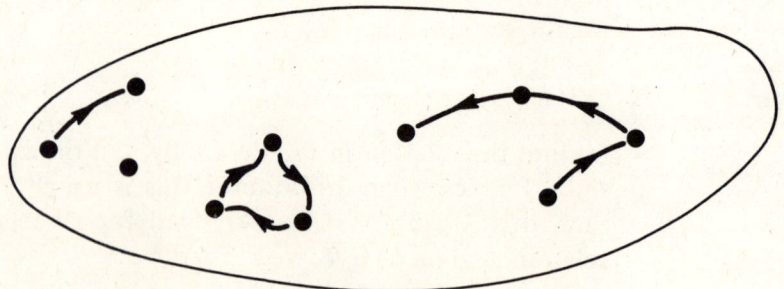

FIGURE 4.6

Complete the graph by drawing the necessary arrows.

PROBLEM 14   Another way to pictorially represent a relation is to graph the relation using Cartesian coordinates. Consider the set $T = \{ 1, 2, 3, 4, 5, 6 \}$ and the relation $\geqslant$ (*is greater than or equal to*). Obviously, $6 \geqslant 1$. Hence, the ordered pair (6, 1) is represented on our graph. Complete the following graph so that all elements of the relation $\geqslant$ on T are plotted.

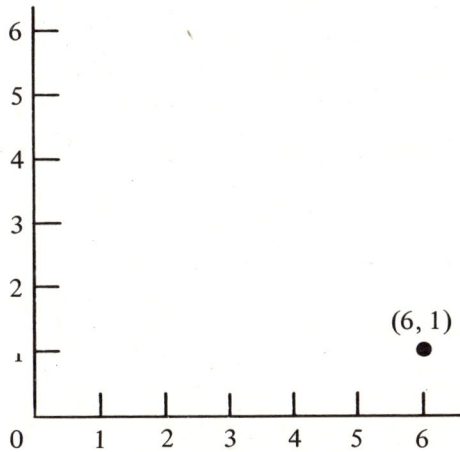

FIGURE 4.7

You now have a definition for a relation defined on a set. You have looked at relations defined on sets of people and on sets of numbers. Actually you have five ways to represent a relation defined on a set: a listing of related elements, a set of ordered pairs, a criterion stated in words or symbols, a directed graph, and a Cartesian coordinate graph.

PROBLEM 15   Take one of the simplest of all relations $\{ (x, y) \mid x = y \}$ as defined on the set of integers between 3 and 8. List the ordered pairs belonging to this relation. List the related elements using the usual symbol for the equal relation. State the criterion for this relation in words.
Draw a directed graph for this relation.
Draw a Cartesian coordinate graph for this relation.

4.2 PROPERTIES OF RELATIONS

When you studied operations in Chapter 2, one of the important points of interest was the examination of some properties that a number of operations had in common. Looking at the properties provided a systematic way to examine the variety of ways in which sets are structured by operations. The same thing can be done with relations to examine ways in which sets are structured by them. In Section 4.1 several properties of relations are explored, but not named. Three properties of relations are the reflexive, symmetric, and transitive properties.

> **DEFINITION:** A relation R defined on a set S is said to be *reflexive* if and only if for each $a \in s$, $a\ R\ a$. (That is, each element is related to itself.)

PROBLEM 1 Apply the definition of the reflexive property to the relations in Problems 5, 7, and 13 in Section 4.1.

PROBLEM 2 Suppose $S = \{1, 2, 3, 4\}$ and a relation $R = \{(1, 1), (2, 2), (4, 4)\}$. Is R reflexive? Why or why not?

PROBLEM 3 It is rather easy to determine if a relation is reflexive by looking at a directed graph for the relation. Decide how you would do this and draw a directed graph for a reflexive relation.

The graph of the relation, *is a sister of,* on a set of girls has the property that whenever an arrow goes from point a to point b, there must also be an arrow from point b to point a. This property is called the symmetric property.

PROBLEM 4 Draw a directed graph for a symmetric relation.

> **DEFINITION:** A relation R defined on a set S is said to be *symmetric* if and only if $a\ R\ b$ implies that $b\ R\ a$. (That is, whenever a is related to b, b must also be related to a.)

124

PROBLEM 5   Is the relation, *is a sister of,* on a set of boys and girls a symmetric relation? Why or why not?

PROBLEM 6   Find two of the previous relations that are symmetric and two that are not symmetric. Explain your choices.

The relation $<$ has the property that if $a < b$ and $b < c$, then it must also be true that $a < c$. This property is called the transitive property.

PROBLEM 7   Draw a directed graph for a transitive relation.

DEFINITION: A relation R defined on a set S is said to be *transitive* if and only if $a\ R\ b$ and $b\ R\ c$ implies $a\ R\ c$. (That is, whenever a is related to b and b is related to c, then it must be true that a is related to c.) Note a, b, and c are not necessarily different elements.

PROBLEM 8   Suppose $S = \{1, 2, 3\}$ and a relation on S is $R = \{(1, 2), (2, 3), (1, 3)\}$. Is R transitive? Why or why not?

PROBLEM 9   In the above definitions for the transitive property, don't forget to consider the cases where elements a, b, and c are not all different elements. For example, if $S = \{1, 2, 3\}$ and $R = \{(1, 2), (2, 1)\}$, is R transitive? Why or why not?

PROBLEM 10   Let $S = \{p, q, r, t\}$.
   a.   Give at least two relations on set S which are reflexive.
   b.   Give at least two relations on set S which are symmetric.
   c.   Give a relation on set S which is transitive, but not reflexive.
   d.   Give a relation on set S which is transitive and symmetric, but not reflexive.

PROBLEM 11   Find two of the previous relations that are transitive and two that are not transitive. Explain your choices.

In this section you have taken specific definitions for properties of relations and applied them. This is a deductive process of testing whether an example fits the definition.

HOMEWORK EXERCISES

Let $S = \{1, 2, 3, 4, 5\}$.

1. Choose relations on set S which illustrate that the 3 properties (reflexive, symmetric, transitive) are independent. (That is, a relation that is *only* reflexive, a relation that is *only* symmetric, and a relation that is *only* transitive.)

2. Give a relation on S which satisfies all 3 properties.

4.3   RELATIONS AND OPERATIONS

The symmetric property of a relation is sometimes confused with the commutative property of an operation. However, the basic confusion is probably in the concepts of relation and operation rather than in their respective properties. While the concepts are not totally independent, they are different. Saying that *m is 2 more than b* is an example of a relation comparing *m* and *b* with the criterion, *is 2 more than*. The equation, $2 + b = m$, is an example of an operation which assigns *m* to the pair 2 and *b* under addition. The first expression compares two elements *m* and *b* of a set; the second expression assigns a third element *m* to two elements, 2 and *b*, of a set.

PROBLEM 1   In the following list write R next to the expressions that represent relations and O next to those that represent operations.
   a. a is a multiple of b, $a, b \in Z$ _____
   b. $a \# x = c$, $a, x, c \in W$. (The operation, #, is defined by the problems in Section 3.1.) _____
   c. $4 @ d = \overset{.}{4}, d \in N$ _____
   d. a is equal to b _____
   e. z divided by m is t, $z, m, t \in Q$, $m \neq 0$ _____
   f. $a > b$, $a, b \in W$. _____
   g. $R = \{(1, 2), (1, 3), (2, 4), (55, 55), (2, 1), (3, 1), (4, 2)\}$ _____
   h. _____ (Refer to Table 4.1.)

TABLE 4.1

*	1	2	3	4
1	1	2	3	4
2	4	3	2	1
3	3	4	1	2
4	2	3	4	1

The relation R in **g** is shown as a set of ordered pairs. All of the other relations can also be written in this form. In fact, as you recall, a relation is defined as a set of ordered pairs.

When the set upon which the relation is defined is finite and small, every ordered pair belonging to the relation can be listed. If the set is finite and large or infinite, only some of the ordered pairs can be shown or the set builder notation must be employed. $R = \{(x, y) | x > y, x, y \in W\}$ is one way of writing the relation defined by the criterion, *is greater than,* on the set of whole numbers.

PROBLEM 2   Put several relations from the list in Problem 1 in the form of ordered pairs.

The operation * in part **h** of Problem 1 is given in the form of a table. All of the operations can be written as tables also, though, as with relations, when the set is either finite and very large or infinite, only a small portion can be displayed.

Table 4.2 is an illustration of the addition table for whole numbers.

PROBLEM 3   Put several operations from the list in Problem 1 in the form of a table.

Now that the distinction between operations and relations has been explored, you can look at the symmetric property of relations and the commutative property of operations.

The easiest way to test a relation to see if it is symmetric is to examine the ordered pairs. If every time the ordered pair

TABLE 4.2

+	0	1	2	...
0	0	1	2	...
1	1	2	3	..
2	2	3	4	..
3	3	4	.	..
4	4	.	.	..
5	.	.		
:				

(*a*, *b*) is in the relation the ordered pair (*b*, *a*) also appears, then the relation is symmetric. Let R be the relation, *is the mother of*, defined on the set of all women in Maryland. If *a* is the mother of *b*, then (*a*, *b*) belongs to R, but (*b*, *a*) doesn't because *b* is not the mother of *a*. Therefore we conclude that *is the mother of* is not a symmetric relation.

The easiest way to test an operation to see if it is commutative is to examine the table. If a diagonal line drawn from the upper left hand corner of the table to the lower right hand corner divides the table into two halves that are mirror images of one another, then the operation is commutative. Table 4.3 is a partial table for the operation $a \gamma b = a + 2b$, $a, b \in W$.

TABLE 4.3

γ	0	1	2	3	.	.	.
0	0	2	4	6	.	.	
1	1	3	5	7	.		
2	2	4	6	8	.		
3	3	5	7	9	.	.	
.	.	.					

The diagonal line shows that this operation is not commutative because the half of the table below the line is clearly not a mirror image of the half above the line. Another way to look at commutativity is to check and see that if the ordered pair (a, b) is assigned to c by the operation then the ordered pair (b, a) will also be assigned to c by the operation.

PROBLEM 4   List the relations from Problem 1 which are symmetric.

PROBLEM 5   List the operations in Problem 1 which are commutative.

4.4 EQUIVALENCE RELATIONS AND EQUIVALENCE CLASSES

Let $S = \{0, 1, 2, 3, \ldots, 18\}$. Consider a relation R, defined on S as follows: $a R b$ if and only if a and b differ by a multiple of 3. For example:

$$4 R 10 \text{ because } 10 - 4 = 6 \text{ and } 6 \text{ is a multiple of } 3$$

$$7 R 7 \text{ because } 7 - 7 = 0 \text{ and } 0 \text{ is a multiple of } 3$$

$$7 R 4 \text{ because } 7 - 4 = 3 \text{ and } 3 \text{ is a multiple of } 3$$

PROBLEM 1   Find several examples to illustrate that the relation R as defined is:
a. reflexive,
b. symmetric,
c. transitive.

PROBLEM 2   List the elements of at least the following first six sets using the relation R defined at the beginning of this section.

$\{x \mid 1 R x, x \in S\}$ is read as the set of all x such that 1 is related to x. The first two elements of S_1 are 1 and 4.

$$S_1 = \{ x \mid 1 R x, x \in S \} = \{ 1, 4 \qquad\qquad\qquad \}$$

$$S_2 = \{ x \mid 2 R x, x \in S \} = \{ \qquad\qquad\qquad\qquad \}$$

$$S_3 = \{ x \mid 3 R x, x \in S \} = \{ \qquad\qquad\qquad\qquad \}$$

$$S_4 = \{ x \mid 4 R x, x \in S \} = \{ \qquad\qquad\qquad\qquad \}$$

$$S_5 = \{ x \mid 5 R x, x \in S \} = \{ \qquad\qquad\qquad\qquad \}$$

$$S_6 = \{ x \mid 6 R x, x \in S \} = \{ \qquad\qquad\qquad\qquad \}$$

$$S_7 = \{ x \mid 7 R x, x \in S \} = \{ \qquad\qquad\qquad\qquad \}$$

.

.

.

$$S_{18} = \{ x \mid 18 R x, x \in S \} = \{ \qquad\qquad\qquad\qquad \}$$

Do you notice a pattern developing? If so, what is it? If not, list the elements of some additional sets. Looking at these sets, how many distinct (different) sets are there?

You should have found 3 distinct sets in Problem 2. These sets are S_1, S_2, and S_3.

PROBLEM 3   Complete the following by filling in the blanks to make true statements using the sets from Problem 2.

a. $S_1 \cap S_2 = $ _____

b. $S_2 \cap S_3 = $ _____

c. $S_1 \cap S_3 = $ _____

d. $S_1 \cup S_2 \cup S_3 = $ _____

e. S_1 _____ $S (=, \subseteq)$

f. S_2 _____ $S (=, \subseteq)$

g. S_3 _____ $S (=, \subseteq)$

PROBLEM 4   Figure 4.8 is a picture of set S and the subsets S_1, S_2 and S_3. Fill in the elements of S_1, S_2 and S_3.

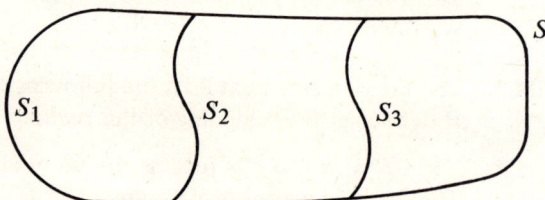

FIGURE 4.8

The relation R, and any other relation that is reflexive, symmetric, and transitive, is called an *equivalence relation.* Any equivalence relation forms subsets (of the set that the relation is defined on) such that each element in a particular subset is related to every other element in that subset. These subsets are called *equivalence classes.*

PROBLEM 5   Do S_1, S_2, and S_3 fit the definition of equivalence classes?

PROBLEM 6   Refer to Problem 2 as necessary.
    a. Is any element of S in more than one equivalence class?
    b. Is there an element of S which does not belong to an equivalence class?
    c. Generalize your answers to *a* and *b* by completing the following: If R is an equivalence relation on S, then any element of S belongs to _____ equivalence class(es) formed by R.

PROBLEM 7   Let X be the relation, *has the same sex as,* defined on the set P of all persons in your math class. Is X an equivalence relation? Tell why or why not. If X is an equivalence relation describe the equivalence classes formed by X.

PROBLEM 8   Let A be the set $\{1, 2, 3, 4, 5, 6, 7, 8, 9\}$ and let M be the relation, *is a multiple of.* Is M an equivalence relation? Tell why or why not. If M is an equivalence relation, describe the equivalence classes formed by M.

PROBLEM 9   Check previous examples of relations and determine which are equivalence relations.

One final comment on equivalence classes. You have been using them since the beginning of this course. The elements of Z_7 are really equivalence classes. Each element can be thought of as an equivalence class formed by a relation defined on a set.

★PROBLEM 10   Consider the elements of Z_7 as equivalence classes. Some relation must form these equivalence classes. Determine what this relation is for Z_7. What is the set on which this relation is defined?

HOMEWORK EXERCISES

Decide whether the following relations are equivalence relations. If so, describe the equivalence classes. If not, tell which property is not satisfied and give an example.

1. $\{0, 1, 2, \ldots, 20\}$ *differs by a multiple of 5 from* . . . .

2. $\{$all students in your math class$\}$ . . . *is in the same group as* . . . .

3. $\{0, 1, 2, \ldots\}$ ... *is 5 more than* ....

4. S is the set of rational numbers. $a \, R \, b$ if $a = b$ or $a = \bar{}b$.

5. S is the set of all people in the world today. If a lives within 100 miles of b, $a \, R \, b$.

The set of rational numbers, Q, can be defined in terms of equivalence classes. Each rational number is an equivalence class which is a set of ordered pairs of integers Z. Note that this is a bit more complicated than other relations since the elements of the set are ordered pairs. The relation R defined on these ordered pairs of Z is not too complicated. If a, b, c, $d \in Z$, then $(a, b) \, R \, (c, d)$ if and only if $a \cdot d = b \cdot c$, $b \neq 0$ and $d \neq 0$. For example $(1, 2) \, R$ $(4, 8)$ since $1 \cdot 8 = 2 \cdot 4$.

PROBLEM 11   Find four ordered pairs of integers that are related to $(3, 4)$ by the relation R as defined.

All the ordered pairs of integers that are related to $(3, 4)$ by the relation R belong to one equivalence class. This equivalence class is a rational number which can be named 0.75, $\dfrac{3}{4}$, $\dfrac{12}{16}$ ...

PROBLEM 12   Find four ordered pairs of integers that name the rational number 1.5.

Find four ordered pairs of integers that name the rational number $\bar{}2.2$.

4.5 FINITE AFFINE GEOMETRY

When studying infinite algebraic systems it was helpful to consider the finite systems $(Z_n, +, \cdot)$ in order to focus more clearly on the ways in which the operations of addition and multiplication structure a set of numbers. The study of finite geometries can provide insight into the more familiar infinite geometries. Relations such as parallel and collinear will be examined as they structure a set of points. Furthermore, this investigation of finite geometries illustrates and allows for practicing mathematical processes such as formulating conjectures, proving theorems, and writing and testing definitions.

First it is necessary to define a finite geometry that you can explore. Start with the following clues with which you can develop a finite geometric structure.

A box contains a number of beads and strings. The exact number of beads and strings is unknown, but some clues are given. Use the information contained in the clues to determine the number of beads and the number of strings in the box. There may be more than one correct solution.

Clue 1.   There are at least two beads in the box.

Clue 2.   Every two beads belong to one and only one string.

Clue 3.   Not all beads are on the same string.

Clue 4.   Every string has at least two beads on it.

PROBLEM 1   Some suggested solutions are illustrated in Figure 4.9, but they are not necessarily correct. Decide which ones, if any, are correct. If a solution is not correct, explain why not.

Proposed solution 1   Proposed solution 2

Proposed solution 3

FIGURE 4.9

PROBLEM 2   Clue 1 says there are *at least* two beads in the box. Does this mean there could be more than two beads? Could it mean there are exactly two beads?

While Clue 1 allows the possibility that there are exactly two beads, the four clues taken together do *not* allow exactly two beads. Here is the proof that there are at least three beads in the box based on the clues.

> ***THEOREM:*** There are at least three beads in the box. By Clue 1, there are at least two beads; call them *A* and *B*. By Clue 2, there is a string containing both *A* and *B*. Clue 3 states that not all beads are on the same string, so there must be a third bead that is not on the string containing both *A* and *B*.

PROBLEM 3   None of the clues tells how many strings are in the box, but the four clues taken together require at least three strings in the box. Prove the following theorem.

> ***THEOREM:*** There are at least three strings in the box.

PROBLEM 4   Illustrate a three-bead model that satisfies all four clues.

PROBLEM 5   Illustrate a four-bead model that satisfies the four clues.

The puzzle you have been solving concerned beads and strings, but no use is made of any characteristics of beads or strings. According to the clues, a string can be thought of as a set of two or more beads, and the beads can be thought of simply as elements of sets that obey the rules given in the clues. In fact, any words that suggest sets and elements of sets can be substituted for *beads, strings,* and *box.* For example, the words *members, committees,* and *club* could be substituted for *beads, strings,* and *box.* Or nonsense words like *pims, lims,* and *flam* could be used.

PROBLEM 6   Just to see how it works, substitute the words *members, committees,* and *club* for *beads, strings,* and *box.* The first theorem proved becomes: "There are at least three members in the club." Translate the theorem in Problem 3. Are both these statements true?

PROBLEM 7   Write the four clues substituting the words *points, lines,* and *plane* for *beads, strings,* and *box.*

Since points, lines and planes are the subject matter of geometry, the four clues can be taken as the axioms for a geometry. The words, *a geometry,* are intended to suggest that there can be many different geometries, depending on which axioms are used. In this geometry, the word *lines* does not necessarily refer to the continuous straight lines of familiar Euclidean geometry.

A Finite Geometry

Here again are the four basic axioms:

AXIOM 1:   There are at least two points in the plane.

AXIOM 2:   Every two points belong to one and only one line.

AXIOM 3:   Not all points are on the same line.

AXIOM 4:   Every line has at least two points on it.

Now add the following axioms:

AXIOM 5:   Every line contains exactly three points. (**Note:** This axiom replaces Axiom 4.)

AXIOM 6:   Every two lines have a point in common.

Since this new set of six axioms includes the original four axioms, the two statements already proved still hold, namely, that there are at least three points in the plane, and there are at least three lines in the plane.

PROBLEM 8   Test each of the models in Figure 4.10 to see if the model satisfies Axioms 1 to 6. If you find that these models violate your idea of points and lines, simply think of the models and axioms in terms of beads and strings.

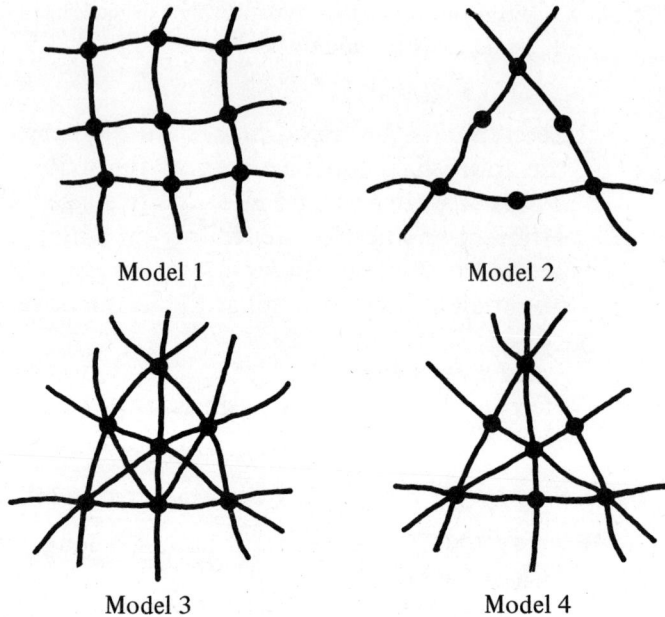

Model 1   Model 2

Model 3   Model 4

FIGURE 4.10

PROBLEM 9   Determine how many points and lines there are in this geometry satisfying Axioms 1 to 6. Hint: Use letters A, B, C, . . . , to represent points and consider each line to be a set of three points; $\{A, B, C\}$ is one line of the plane.

PROBLEM 10   Use the model from Problem 8 satisfying Axioms 1 to 6 to test the following conjectures. This model is unique. Since this is true, you can use this model for proving conjectures. If there had been more than one model, as is the case of those models satisfying Axioms 1 to 4, you would have to test *all* possible models in order to prove a conjecture from models.

THEOREM 1:   There are at least five points on the plane.

THEOREM 2:   Two lines cannot have more than one point in common.

THEOREM 3:   Every point is on at least two lines.

PROBLEM 11   Use the six axioms to prove the three theorems given in Problem 10.

In this section you have repeated familiar processes. First you defined a geometry by assuming a given set of axioms. This was similar to defining a group by a set of properties. One process used was testing models to see if they satisfy given axioms; this is like testing whether a given operation table defines a group. Also, you proved theorems about a finite geometry from the given axioms; in a similar fashion you have proved theorems about groups from group properties.

HOMEWORK EXERCISES

1. Test whether or not the following four models satisfy Axioms 1 to 4. If the model does not satisfy Axioms 1 to 4, indicate which axioms are not satisfied.

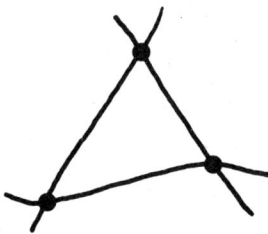

Model 1

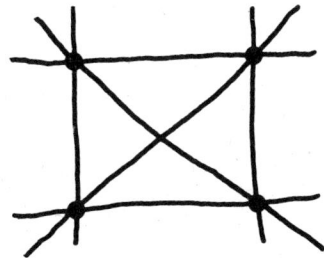

Model 2

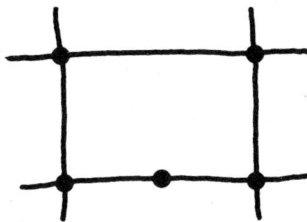

Model 3

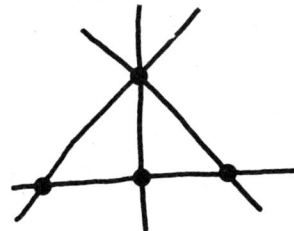

Model 4

FIGURE 4.11

2. For any model in Exercise 1 that does not satisfy Axioms 1 to 4, draw in additional points or lines so that it will satisfy the axioms.

137

3. Using Axioms 1 to 6 prove the following theorem: There are at least two lines that do not contain a given point.

4.6 PARALLEL LINES

Figure 4.12 represents the familiar notion of *parallel lines* in Euclidean geometry. Following is a convenient way of defining parallel lines.

FIGURE 4.12

DEFINITION: Two distinct lines are *parallel* if they have no point in common.

PROBLEM 1   What is the meaning of the word *distinct* as it is used in this definition? Notice that the definition is not applicable if two lines are not distinct.

PROBLEM 2   Apply the definition of parallel lines to the finite geometry in Section 4.5. Does the model that satisfies Axioms 1 to 6 contain distinct parallel lines? If yes, name a pair of parallel lines. If no, explain why not.

PROBLEM 3   Is there a model which satisfies only Axioms 1 to 4 and contains parallel lines? If yes, illustrate one and name a pair of parallel lines. If no, explain why not.

A Finite Geometry with Parallel Lines

Here is a set of axioms which is the same as Axioms 1 to 6, with one exception. Axiom 6 has been replaced by Axiom 6', which guarantees parallel lines.

AXIOM 1:   There are at least two points in the plane.

AXIOM 2:   Every two points belong to one and only one line.

AXIOM 3: Not all points are on the same line.

AXIOM 4: Every line has at least two points on it. (May be omitted since it is replaced by Axiom 5.)

AXIOM 5: There are exactly three points on a line.

AXIOM 6: Through a point not on a given line there is exactly one line parallel to the given line

PROBLEM 4 Nine points and 12 lines are needed to make a model for this set of axioms. List the lines; then illustrate the model.

PROBLEM 5 Name a pair of parallel lines in your model for this new set of axioms. Then list all pairs of parallel lines.

Check the model for Problems 4 and 5 with your instructor.

PROBLEM 6 Following is a list of conjectures about this new geometry. Decide which are true and which are false. Show a counterexample for those which are false.
   a. Two lines cannot have more than one point in common.
   b. Every point is on at least two lines.
   c. Two distinct lines parallel to a third line are parallel to each other.
   d. Every two lines have a point in common.
   e. Every point is on exactly three lines.
   f. If *X* is a point not on line *l,* then there is a line containing *X* which has no point in common with line *l.*

HOMEWORK EXERCISES

Explore the geometry defined by the following relationships of lines and points.

Axiom 1: There are at least two points in the plane.

Axiom 2: Two distinct points belong to exactly one line.

Axiom 3: Not all points are on the same line.

Axiom 4: If point *X* does not lie on line *l,* there is exactly one line containing *X* and not meeting *l.*

Axiom 5: Every line contains exactly two points.

1. Illustrate a model for the axioms.

2. Are there any distinct parallel lines? If yes, list them. If not, explain why not.

3. Examine the following conjectures. Decide which are true and which are false for the model you developed.

   Conjecture 1. Every point belongs to at least two lines.
   Conjecture 2. Every pair of lines has a point in common.
   Conjecture 3. Every point belongs to exactly three lines.

4. Provide a counterexample for each conjecture you believe is false. Write a proof based on Axioms 1 to 5 for each conjecture you believe is true.

4.7 GEOMETRIC RELATIONS

You have now explored finite geometries with and without parallel lines. Defining parallel lines enabled you to examine the geometries in more detail. The parallel relation is an important relation in geometry; points and lines are structured by this relation.

PROBLEM 1   Look at the parallel relation as it is defined on the set of lines in the finite geometry of Section 4.6. The rule to determine if two lines are related is simply: *is parallel to.*

Line *l* is related to line *m* if and only if *l* is parallel to *m* (*l* ∥ *m*).

Is this relation an equivalence relation? Recall that an equivalence relation is reflexive, symmetric, and transitive.

PROBLEM 2   You have found that the relation defined by the rule, *is parallel to,* on the set of all lines in the finite geometry of Section 4.6 was not an equivalence relation. Formulate a precise definition of parallel so that this does turn out to be an equivalence relation. Check your definition with your instructor.

PROBLEM 3   a. Describe the equivalence classes for the definition of parallel.
b. How many are there in the finite geometry of Section 4.6?
c. List the lines in each class.

PROBLEM 4   You have just explored the relation parallel for the set of all lines in a finite geometry. Now, examine the same relation in Euclidean geometry. Consider the set of all lines in a fixed plane. The blackboard is a model of a fixed plane. Use your intuitive notion of planes, lines, and points. Is the parallel relation an equivalence relation?

PROBLEM 5   Re-examine your definition of parallel and write a precise definition of parallel so that it turns out to be an equivalence relation. Check your definition with your instructor.

In space, it is also interesting to explore another relation defined on the set of lines, *is skew to.*

DEFINITION: Line *l* is *skew* to line *m* if and only if *l* and *m* are non-intersecting and do not lie in the same plane.

PROBLEM 6   Investigate the properties of the skew relation. Is this relation an equivalence relation?

HOMEWORK EXERCISES

1. Let the parallel relation be defined on the set of all lines in space. Write a definition of the relation. Is it an equivalence relation?

2. Let the parallel relation be defined on the set of all planes in space. Write a definition of the relation. Is it an equivalence relation?

PROBLEM 7   Now consider the relation defined by the rule, *is perpendicular to,* defined on the set of all lines on a fixed plane. Write a definition by completing

Let α be a fixed plane

Let l and m be any lines on α

l is perpendicular to m $(l \perp m)$ if and only if

Is this relation an equivalence relation?

PROBLEM 8   a. In your definition for perpendicular on the set of all lines on a fixed plane, is there any sensible modification of your definition so that this might become an equivalence relation?

b. Notice, in particular, the transitive property: Let l, m, n be lines on your plane. If $l \perp m$ and $m \perp n$, what conclusion *must* you draw about l and n?

PROBLEM 9   Now explore the perpendicular relation defined on the set of all planes in space.

a. Is this relation an equivalence relation?

b. Again explore, in particular, the transitive property: Let α, β, δ be planes in space. If $\alpha \perp \beta$, and $\beta \perp \delta$, what kinds of conclusions can you draw about α and δ? Be careful, the answer is different from the situation of lines in a plane. Use the cardboard bottle dividers to help you visualize this.

There are more geometric relations. Investigate the relation defined by the rule, *is collinear with,* defined on the set of all points on a plane.

DEFINITION: Point P is collinear with point Q if and only if P lies on the same line as Q.

PROBLEM 10   Investigate the properties of the collinear relation. Is it an equivalence relation?

PROBLEM 11   If the collinear relation is an equivalence relation, describe the equivalence classes. Something unusual happens. Can you explain why?

PROBLEM 12   **a.** If you were to make a checklist of all geometric relations that we have described and their properties, you would find that they are all symmetric. Why do you think this is?

       b. Can you think of a geometric relation that is not symmetric? Check your ideas with your instructor.

HOMEWORK EXERCISES

3. Consider the perpendicular relation defined on the set of all lines in space.
   a. Investigate the properties of this relation.
   b. If $l \perp m$ and $m \perp n$, what conclusions can you draw about l and n?

   Compare this situation with that which arises in Problem 8 of this section.

4. Explore the relation described by the rule, *is equal to,* defined on the set of all points on a plane.
   a. Investigate the properties of this relation.
   b. Examine the equivalence classes if they exist.

   Compare this situation with that which arises in Problem 11 of this section.

5. Explore the *coplanar* relation defined on the set of all lines in space.

DEFINITION: l is coplanar with m if and only if l lies on the same plane as m.

   a. Investigate the properties of this relation.
   b. Examine the equivalence classes if they exist.

   Compare this situation with that which arises in Problem 11 of this section.

6. Following is one definition of the parallel relations that you may have used on the set of lines on a fixed plane.

DEFINITION: Let α be a fixed plane and l and m be any lines in that plane; l is parallel to m if and only if l and m do not intersect or l and m coincide.

This can be shortened by using familiar symbols—especially those for sets.

> ***DEFINITION:*** Let $l, m \in \alpha$. $l \parallel m$ iff $l \cap m = \{\ \}$ or $l = m$.

It is convenient to use these symbols. Write the definition for the parallel and skew relations for lines in space.

7. Make a list of the definitions used in this section for your reference.

4.8   A SPECIAL RELATION

In the next several sections, factors and divisibility will be explored. Through one relation a rich collection of mathematics is developed. There will be opportunities to once again formulate definitions, make conjectures, and test them. The special relation is depicted in the following directed graph.

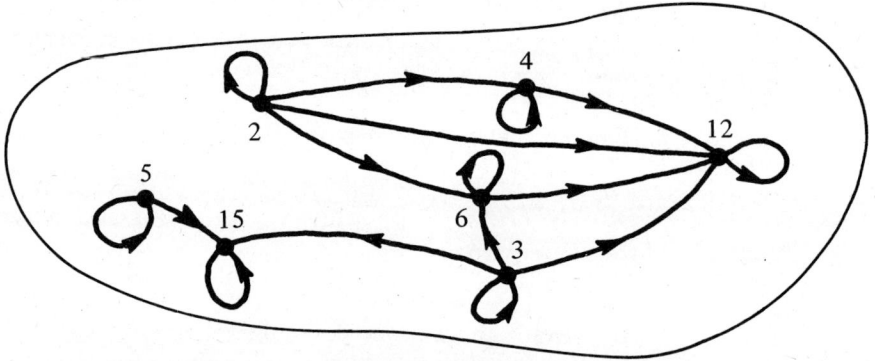

FIGURE 4.13

PROBLEM 1   Study the directed graph in Figure 4.13 and determine the properties of the relation represented by this graph.
  a. Is it reflexive?
  b. symmetric?
  c. transitive?

PROBLEM 2   a. From the information in Figure 4.13 determine the relation pattern the directed graph depicts.

144

b. In Figure 4.14 try to complete the directed graph using the same relation pattern as in Figure 4.13.

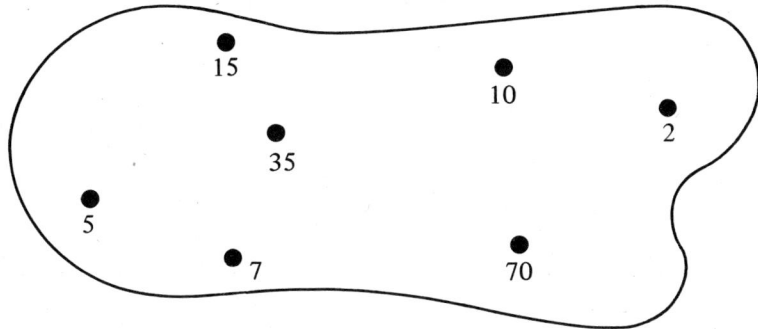

FIGURE 4.14

PROBLEM 3   Using the graph in Figure 4.14 provide two specific examples that illustrate each of the properties of this relation.

PROBLEM 4   The graph in Figure 4.15 is supposed to represent the relation pattern in Figures 4.13 and 4.14. What additional arrows, if any, are needed?

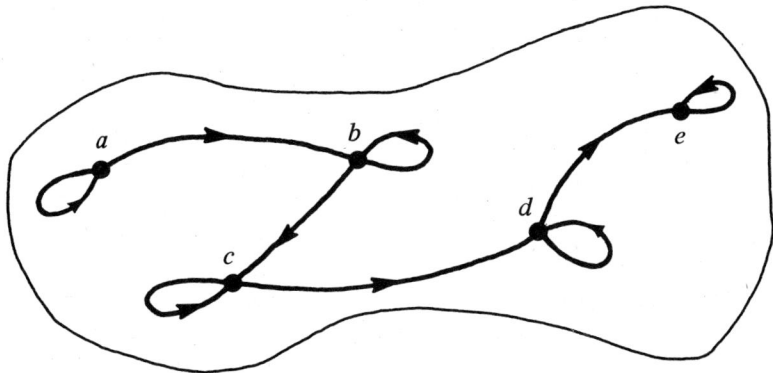

FIGURE 4.15

PROBLEM 5   Using the three graphs in Figures 4.13, 4.14, and 4.15 as a guide, try to define this relation by completing the following with an algebraic ex-

pression: for every *a, b* in *Z, a* is a *factor* of *b* if and only if _____
_____. Be sure to check this definition with the instructor.

The notation *a* | *b* is used to indicate that *a* and *b* are related by the relation, *is a factor of.* For example 5 | 5, 3 | 27, and 8 | 16.

The notation *a* ∤ *b* is used to indicate that *a* is not a factor of *b.* For example 5 ∤ 4, 3 ∤ 10, and 8 ∤ 15. From the notation *a* | *b* both of the following are true: *a* is a factor of *b,* and *b* is a multiple of *a.* 3 | 27 indicates both that 3 is a factor of 27 and 27 is a multiple of 3.

The notation *a* | *b* is sometimes read *a divides b.* However, *a* | *b* should not be confused with the notation of a fraction, *a/b* (read *a divided by b*). The phrase *divided by* indicates an operation and can be checked to see if the commutative or associative properties are met. The term *divides* is a relation and can be examined to see if it satisfies the reflexive, symmetric, or transitive properties of a relation.

PROBLEM 6   Using your definition for the relation, *is a factor of,* determine which of the following are true all the time.

a. 2 | 4   k. 1 | 1
b. 3 | 36   l. 8 | 52
c. 5 | 14   m. 39 | 9
d. 3 | 0   n. 3 | 21
e. 3 | ⁻24   o. ⁻1 | 1
f. 0 | 5   p. If *a* | 30 then *a* is even
g. ⁻4 | 28   q. If *a* | 5 then *a* | 5*x* + 5
h. 7 | 42   r. If *a* | *b* and *b* | *a* then *a* = *b*
i. ⁻2 | ⁻18   s. If *a* | *b* then *a* | *b* + 2
j. 4 | 52   t. If *a* ∤ *b* then *a* | *b* + 2

Examples **c, f, l, m, p, r, s,** and **t** should be false; the others should be true. Do all of your answers agree with these? If not, revise your definition so that your answers will agree.

PROBLEM 7   Now consider the case 0 | 0. According to your definition, is this true? Mathematicians have generally agreed that this should *not* be true and have defined the relation, *is a factor of,* accordingly. If necessary, revise your definition (this will be the last time!) so that 0 | 0 is not true. (Hint: How many integers can you multiply times 0 to get 0?) Check this definition with your instructor.

The following two articles offer helpful suggestions for dealing with the problem of zero in the classroom: (1) Charles Brumfiel, "Zero Is Highly Overrated," *Arithmetic Teacher,* May 1967; (2) Hilda Duncan, "Division by Zero," *Arithmetic Teacher,* October 1971.

HOMEWORK EXERCISES

1. Construct a directed graph for the relation, *is a factor of,* using the following numbers: 2, 3, 7, 12, 15, 21, 42, 50, 75.

2. The directed graph in Figure 4.16 represents the relation, *is a factor of.*

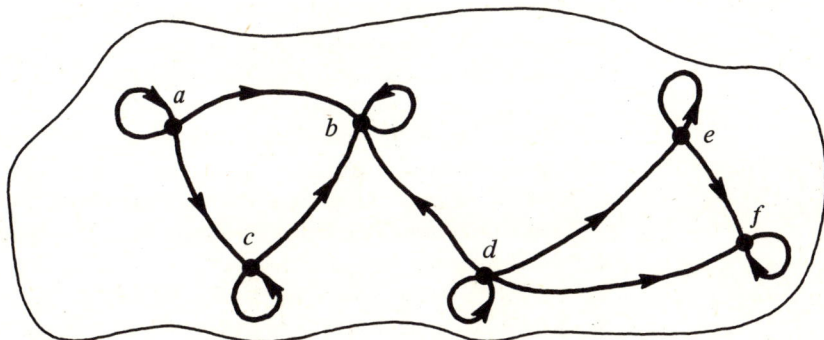

FIGURE 4.16

Find 2 different sets of numbers that can be used as replacements for *a, b, c, d, e,* and *f.*

3. Using your definition for the relation, *is a factor of,* show that each of the following satisfy that definition.
   a. 3 | ⁻15
   b. 23 | 69
   c. ⁻17 | ⁻102
   d. ⁻12 | 132

For your future reference, the final definition of the relation, *is a factor of,* follows.

DEFINITION: For every a, $b \in Z$, a is a factor of b if and only if there exists a unique integer k such that $b = a \cdot k$.

4.9 A LITTLE BIT OF LOGIC

The relation defined by the rule, *is a factor of,* is an important part of mathematics known as number theory. The modular arithmetic that you have studied is also considered part of number theory.

PROBLEM 1 There are many theorems that depend on an understanding of the definition of the relation, *is a factor of.* Consider the following four statements. If you can justify a statement, write *true* in the blank. If the statement is false, write *false,* and indicate why.

Note those cases where the variable cannot be zero. Recall that if a statement contains the phrase, "for every a, b," then the statement can be true only if it is true for all allowable values in the given set. If a statement contains the phrase, "there exists an a, b," then the statement is true if there is at least one pair of values for a and b in the given set that makes the statement true.

a. For every a, if $a \mid 60$ then a is a multiple of 5. _____

b. There exists an a that is a multiple of 5 such that $a \mid 60$. _____

c. There exists an a that is a multiple of 7 such that $a \mid 60$. _____

d. There exists an a that is even such that $a \mid 60$. _____

In Problem 1 statements **a** and **c** are false. Statement **a** is false because $12 \mid 60$, and **c** is false because there is no multiple of 7 that is a factor of 60. In none of the cases was the variable a ever zero.

PROBLEM 2 Continue your work by examining the following problems and determining if they are true or false. Justify each false answer.

a. For every $a \in Z$, $1 \mid a$. _____

b. For every $a \in Z$, $a \mid 1$. _____

c. For every $a \in Z - \{0\}$, $a \mid a$. _____

d. There exists an $a \in Z$, such that $a \mid a$. _____

e. If $a \mid b$ and $a \mid c$, then $a \mid bc$ for every a, b, $c \in Z$. _____

f. If $a \mid bc$, then $a \mid b$ and $a \mid c$ for every $a, b, c \in Z$. _____

g. There exists $a, b, c \in Z$ such that if $a \mid bc$, then $a \mid b$ and $a \mid c$.

_____

h. If $a \mid b$ and $b \mid a$, then $a = b$ for every $a, b \in Z$. _____

i. If $a = b$, then $a \mid b$ and $b \mid a$, for every $a, b \in Z$, $a \neq 0, b \neq 0$.

j. If $a \mid (b + c)$, then $a \mid bc$ for every $a, b, c \in Z$.

k. For every $a, b, c \in Z$, if $a \mid bc$ then $a \mid (b + c)$.

l. If $a \mid b$ and $a \mid c$, then $a \mid (b + c)$ for all $a, b, c \in Z$.

m. If $a \mid (b + c)$, then $a \mid b$ and $a \mid c$ for all $a, b, c \in Z$.

n. If $a \mid (b + c)$ and $a \mid b$, then $a \mid c$ for all $a, b, c \in Z$.

o. If $a \mid b$, then $a \mid {}^{-}b$ for all $a, b \in Z$.

p. If $a \mid {}^{-}b$, then $a \mid b$ for all $a, b \in Z$.

q. There exists $a, b, c \in Z$ such that $a \mid (b + c)$ and $a \mid bc$.

Almost all of the statements in Problem 2 are written in the same form. It is particularly useful for us to study the logic of these types of statements since almost every statement in mathematics is in this form or can be put in this form. Any statement of the form: If _____, then _____, is called an *implication*. Whatever fills the blank following if, is called the *antecedent*; the phrase that follows then is called the *consequent*. For example, the antecedent of part e of Problem 2 is, "$a \mid b$ and $a \mid c$"; the consequent is "$a \mid bc$." In mathematical symbols this would be written: $a \mid b$ and $a \mid c \Rightarrow a \mid bc$, where the arrow would be read *implies*, hence the name implication.

PROBLEM 3   Pick five statements out of the list in Problem 2 and label the antecedents and consequents.

PROBLEM 4   Now investigate the relationships between certain implications in Problem 2. Consider the following pairs of statements: **e** and **f**, **h** and **i**, **j** and **k**. Describe the relationship between the statements in each pair. Can you find another pair of statements in the list with this relationship? If so, list the pair _____.

Statements with the relationship investigated in Problem 4 are called *converses*. Simply stated, the converse of an implication is formed by exchanging the antecedent and consequent.

★PROBLEM 5   Looking at the list of statements in Problem 2, answer the following questions:
  a. If a statement is true, is its converse necessarily true? necessarily false?
  b. If a statement is false, is its converse necessarily false? necessarily true?
  c. Is there any definite relationship between the truth values of a statement and its converse? If so, state the relationship. (The *truth value* of a statement is merely whether it is true or false.)

HOMEWORK EXERCISES

1. What is the converse of the converse of a statement?

2. Write the converses of the following statements:
    a. If you play for the Bullets, then you are a pro basketball player.
    b. For every $x \in Z$, if $3x = 9$, then $x = 27$.

3. Write the converse of each statement; then judge its truth value.
    a. If a polygon is a square, then it is a rectangle.
    b. If you sign up for a course under pass-fail grading, then you will not work hard in that subject.
    c. If you disagree with a government policy, then you are un-American.

Often in mathematics we come across statements that contain the phrase, *if and only if.* You may be wondering about the difference between these statements and those containing merely *if.* The statement, "*A* if and only if *B*" (where *A* and *B* represent statements) can be thought of as two separate statements:

A if *B* which can be represented $B \Rightarrow A$ and

A only if *B* which can be represented $A \Rightarrow B$.

These 2 statements are combined symbolically by writing $A \Leftrightarrow B$. Often the phrase, *if and only if,* is abbreviated by *iff.* It should be clear that such an if and only if statement is true when both the if statement is true and the only if statement is true. For the statement, *A* if and only if *B,* to be true, both the statement and the converse formed with *A* and *B* must be true. Symbolically,

$A \Leftrightarrow B$ is true under the conditions where both $A \Rightarrow B$ is true and $B \Rightarrow A$ is true.

PROBLEM 6   All definitions are iff statements. Look at the definition of the relation defined by the rule, *is a factor of*. This is an if and only if statement.

Choose from the list of true-false statements of Problem 2 for the following:
a. Combine statements to produce an example of an iff statement that is true.
b. Combine statements to produce two examples of iff statements that are false.

4.10   PROVING SOME RELATIONS

Up to this point you have merely decided whether statements were true or false. If they were false, you were able to produce a counterexample to back up your claim. The time has come to back up some of the statements you claimed were true. To get you started, fill in the reasons in the following proof:

THEOREM: If $a \mid b$ and $a \mid c$, then $a \mid (b + c)$ where $a, b, c \in Z$.

Proof		*Statements*	*Reasons*
	1.	$a \mid b$ and $a \mid c$	1.
	2.	$b = a \cdot q$ and $c = a \cdot r$ for some $q, r \in Z$	2.
	3.	$b + c = a \cdot q + a \cdot r$	3.
	4.	$b + c = a \cdot (q + r)$	4.
	5.	$a \mid (b + c)$	5.

PROBLEM 1   a. Try this one from scratch!
Theorem: If $a \mid b$ and $a \mid c$, then $a \mid bc$ for all $a, b \in Z$.
b. Now that you're rolling, try these two.
Theorem: If $a \mid b$, then $a \mid {}^-b$ for all $a, b \in Z$.
Theorem: If $a \mid (b + c)$ and $a \mid b$, then $a \mid c$ for some $a, b, c \in Z$.

For this last theorem, did you do a general proof? If so, reread the theorem and decide whether or not there was an easier way to prove the theorem. Actually, all you have to do is find one set of values for *a, b,* and *c* so that the theorem is true. Hopefully, this will point out that it pays to read theorems carefully so that you don't do more work than is necessary.

HOMEWORK EXERCISES

1. Prove: If $a \in Z$, then $1 \mid a$.

2. Prove: There exists $a, b, c \in Z$ such that if $a \mid bc$, then $a \mid b$ and $a \mid c$.

3. Prove: If $a \mid b$, then $a \mid bc$ for every $a, b, c \in Z$.

4.11 FACTORIZATION

You have been delving into some proofs of the relation, *is a factor of.* You will return to them later; now we want to turn our attention to a specific group of numbers.

PROBLEM 1   In the sets of numbers listed in Table 4.4, try to find a rule that describes as closely as possible the relationship between the numbers in each set. Then, list 5 other elements of the set. The first 2 are done to give you a start.

The last rule describing the set *P* is the rule with which we are concerned. Undoubtedly, you are familiar with this set. It is the infinite set of *prime* numbers.

DEFINITION: A number, *p,* is said to be prime if and only if all three of the following conditions are satisfied:

1. $p \in W$;
2. $p > 1$;
3. The only factors of *p* are $1, {}^{-}1, p, {}^{-}p$.

TABLE 4.4

Set	Rule	Elements
$A = \{\ldots, ^-6, ^-4, ^-2, 0, 2, 4, \ldots\}$	For every $x \in A$, $x = 2n$, $n \in Z$	$8, 10, ^-14, 22, ^-44$
$B = \{\ldots, ^-18, ^-11, ^-4, 3, \ldots\}$	For every $x \in B$, $7 \mid (x - 3)$	$^-25, 10, 24, ^-32, 38$
$C = \{\ldots, 4, 8, 12, 16, 20, \ldots\}$		
$D = \{1, 3, 9, 27, 81, 243, \ldots\}$		
$E = \{0, 1, 4, 9, 16, 25, 36, 49, \ldots\}$		
$P = \{2, 3, 5, 7, 11, 13, 17, 19, 23, \ldots\}$		

PROBLEM 2   Compare this definition with the rule that you formulated and determine if there is any difference (i.e., did your definition include any numbers that aren't prime or exclude any that are?). Any natural number greater than 1 that is not prime is called *composite*.

HOMEWORK EXERCISES

Eratosthenes, a Greek who lived in the third century B.C., described the following technique for "sieving" out the composite numbers between 1 and 100. Use Table 4.5 as you perform the nine steps.

TABLE 4.5

1	2	3	4	5	6	7	8	9	10
11	12	13	14	15	16	17	18	19	20
21	22	23	24	25	26	27	28	29	30
31	32	33	34	35	36	37	38	39	40
41	42	43	44	45	46	47	48	49	50
51	52	53	54	55	56	57	58	59	60
61	62	63	64	65	66	67	68	69	70
71	72	73	74	75	76	77	78	79	80
81	82	83	84	85	86	87	88	89	90
91	92	93	94	95	96	97	98	99	100

1. Cross out 1.

2. Circle 2; beginning to count with 3, cross out all the other multiples of 2.

3. Circle the smallest number not yet circled or crossed out. Cross out all other multiples of this number. Are some of these numbers already crossed out? Explain why.

4. Is it necessary to cross out multiples of 4? Explain why or why not.

5. Circle the smallest number not yet circled or crossed out. Cross out all other multiples of this number.

6. Is it necessary to cross out multiples of 6? Why?

7. Are you beginning to observe a pattern? For which of the following will you have to cross out multiples: 7, 8, 10, 11?

8. How far must this process continue, before only primes are left?

9. List the primes between 1 and 100.

One of the many uses of prime numbers involves writing all whole numbers, prime and composite, as products of their prime factors. This is called a prime factorization. For relatively small numbers, it is not difficult to merely write their prime factorizations. For example:

a. $24 = 2 \cdot 2 \cdot 2 \cdot 3 = 2^3 \cdot 3$

b. $63 = 3 \cdot 3 \cdot 7 = 3^2 \cdot 7$

c. $105 = 3 \cdot 5 \cdot 7$

d. $29 = 29$

Notations: It is customary to write the prime factors in increasing order and to denote repetition of factors with exponents. Notice that the prime factorization of the prime number 29 is the prime number 29. For large numbers, however, it is often useful to write the numbers as a product of any two of its factors, and then successively break up each of the factors into their prime factorizations. This procedure is accomplished, rather neatly, through the use of *factor trees.*

154

Examples

a. Write the prime factorization of 60.

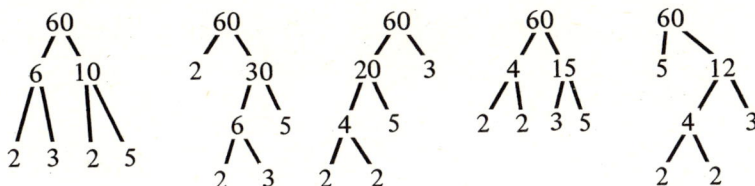

FIGURE 4.17

Thus, no matter which pair of factors you start with, the prime factorization of 60 always turns out to be $2^2 \cdot 3 \cdot 5$.

b. Write the prime factorization of 350.

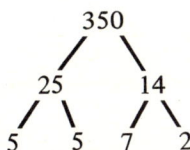

FIGURE 4.18

PROBLEM 3   Find at least three other factor trees for 350 and write them. The prime factorization of 350 is _____ .

The two examples above provide some motivation for one of the basic results in number theory. The theorem states the relationship between prime and composite numbers. It has a rather impressive name which is well deserved as it is crucial to the proofs of numerous theorems.

THE FUNDAMENTAL THEOREM OF ARITHMETIC: Any composite number may be factored into a product of primes. Moreover, this factorization is unique except for order.

The proof of this theorem is much too difficult to be presented here. If you are interested in the proof, see your instructor for a reference. Basically, what the theorem does is to guarantee that if you have found a prime factorization of a number, you have actually found the only prime factorization, regardless of the method you used to obtain it.

HOMEWORK EXERCISES

10. Use factor trees to write the prime factorizations of the following:
    a. 150
    b. 640
    c. 2310
    d. 16170

11. Write the prime factorizations of the following:
    a. 170
    b. 1053
    c. 1295
    d. 2604

Prime factorization is helpful in connection with two operations with which you have already had some experience, namely gcd and lcm.

DEFINITIONS: For all a, $b \in Z$, a gcd b is the largest integer c such that $c \mid a$ and $c \mid b$.
    For all a, $b \in Z$, a lcm b is the smallest positive integer d such that $a \mid d$ and $b \mid d$.

PROBLEM 4   a. Complete Table 4.6.

    b. Using Table 4.6 and concentrating on the prime factorization, describe an algorithm (a step-by-step procedure) that will find the gcd of any two numbers and one which will find the lcm of any two numbers. List the algorithms below.

gcd ALGORITHM:

lcm ALGORITHM:

TABLE 4.6

Number	Prime factorization	a gcd b	Prime factorization of a gcd b	a lcm b	Prime factorization of a lcm b
1 a = 8					
b = 32					
2 a = 84					
b = 90					
3 a = 26					
b = 15					

c. Use your algorithms to find gcd's and lcm's of the following:
   1. 63, 24
   2. 95, 19
   3. 252, 315

HOMEWORK EXERCISES

12. Find gcd and lcm of each pair of numbers:
    a. 170, 264
    b. 56, 39
    c. 98, 47

13. Recall the properties of gcd and lcm and compute the gcd and lcm of each:
    a. 6, 9, 15
    b. 28, 15, 24
    c. 18, 36, 27, 45

14.

DEFINITION: Two numbers, *a, b* are *relatively prime* if and only if *a* gcd *b* = 1.

This means that the only factor that *a* and *b* have in common is one. As an illustration, a fraction is said to be in lowest terms when the numerator and denominator are relatively prime.
    a. List three pairs of numbers that are relatively prime.
    b. If two numbers are relatively prime, then one of the numbers is prime. True or False?
    c. Consider the numbers 5, 49, 31, 6. Are these numbers relatively prime?
    d. Are 2 distinct (different) primes always relatively prime?
    e. If *a* and *b* are relatively prime, what is *a* lcm *b* equal to?

15. a. Figure 4.19 is a graph of the relation, *is a factor of*. Find values for *a, b,* and *c.*

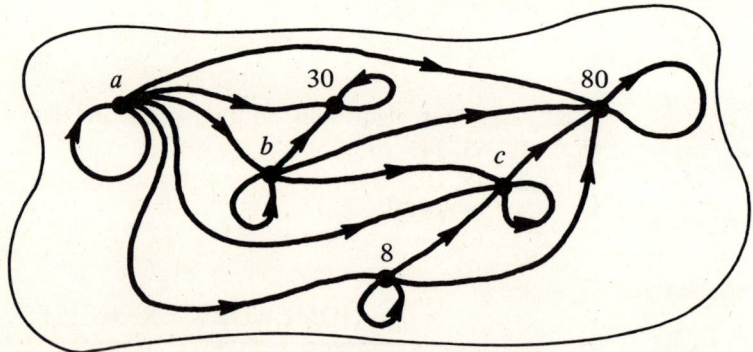

FIGURE 4.19

    b. Pick two sets of five elements such that the elements of the first set are all relatively prime and the elements of the second set are all prime numbers. Construct directed graphs for the relation, *is a factor of,* on each set. Compare the two graphs and decide if there is any difference between the properties illustrated in each graph. Write the difference if there is any. Now compare these graphs to the first two graphs in Section 4.8. Any difference here? If so, what is it?

4.12   HAVE THE COMPUTER DO IT

As you might guess, the computer can be used to good advantage in number theory. For example, programs can be written to find factors (divisors), primes, prime factorizations, greatest common divisors (gcd), and lowest common multiples (lcm).

In the previous work with computer programming in the BASIC language, the INPUT, LET, IF . . . THEN . . . , PRINT, GØ TØ, and END statements were used.

Before introducing some new statements, review the use of some of these statements. An appropriate task for the computer is to have it search for factors. As a review of flowcharting and programming, you will write a computer program to find the factors of any given counting number given the flowchart. The definition for the relation, *is a factor of,* defined on the set N is as follows: For all $a, b \in N$, a is a factor of b if and only if there exists a unique counting number k such that $b = a \cdot k$.

Approach the problem from a very fundamental point of view. Simply take values of a, starting with 1, and try values for k to see if $a \cdot k = b$. Since any factor of b will be equal to or less than b, and k must also be equal to or less than b (otherwise $a \cdot k > b$), stop trying a's and k's when they reach b. Figure 4.20 is one possibility for the flowchart.

PROBLEM 1   Write a BASIC program that matches the flowcharted procedure.

There are a number of programs that could be written to find the factors of a given counting number. The program written by an individual depends upon the individual's knowledge of BASIC and the individual's particular style.

159

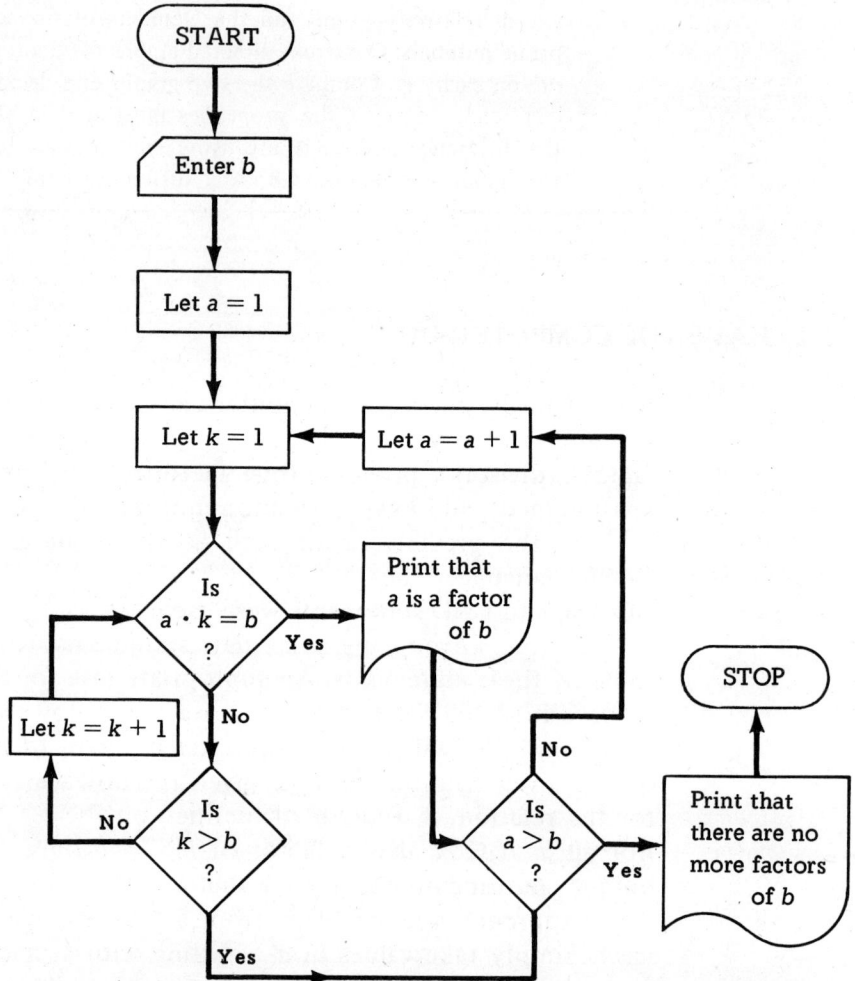

FIGURE 4.20   Flowchart for finding divisors of positive integer.

A feature of the BASIC language we have not yet explored is the existence of special functions built into the language.

For example, there is a square root function denoted SQR. If the programmer uses the statement

```
100 LET X=SQR(Y)
```

The computer automatically computes the square root of the

160

value of Y and puts that value into the storage location called X. Either a constant, a variable, or a valid expression can be used in the parentheses with SQR.

Example:

```
10 LET Y=6.0
20 LET X=SQR(Y)
30 PRINT X
40 END
```

```
        Output
        2.44949
```

Example:

```
100 LET A=7
200 LET B=9
300 LET S=SQR(A+B)
400 PRINT S
500 END
```

```
        Output
        4.00000
```

Another function, which is not built into every version of BASIC, that is especially useful is the MØD function. The MØD function with the following statement is:

```
200 LET X=MØD(Y,M)
```

The computer will put the value of Y (MØD M) into the storage location called X. Again, Y and M can be any constant, variable, or valid expression, but M must be a positive integer.

Example:

```
10 INPUT Y, M
30 LET X=MØD(Y,M)
40 PRINT X
50 END

? 8, 5
3
```

161

Using this function, the program previously written in Section 2.13 for finding the sum and product of numbers in mod 7 could be shortened to the program below.

```
10 INPUT A,B,M
20 LET S=MØD(A+B, M)
30 LET P=MØD(A*B,M)
40 PRINT S, P
50 END
```

Another function which will be of use to us is the INT function. The definition of INT is as follows:

DEFINITION: INT(A) = the largest *integer* equal to or less than A.

Examples:

INT (3.5) = 3
INT (⁻4) = ⁻4
INT (⁻4.2) = ⁻5
INT (7) = 7
INT (5/3) = 1
INT (⁻20/3) = ⁻7

HOMEWORK EXERCISES

1. Find the value of each computer expression below.
   a. MØD (11, 7)
   b. MØD (5 + 3, 7)
   c. MØD (2 * 3, 5)
   d. INT (7/2)
   e. INT (3.1 * 2.5)
   f. INT (6.3/.81)

2. a. Suppose A is a divisor of B (i.e., there exists a unique integer K such that B = A * K).

   What is the value of INT (B/A)? What is the value of B/A?

   b. Suppose A < B and A ∤ B. What can you say about INT (B/A) compared to B/A? Justify your response with some examples.

3. Rewrite the program for finding additive and multiplicative inverses in mod 7 using the MØD function.

4. Write a program for printing out the addition and multiplication facts in mod 5. Write your program so the output looks like:

MØD 5 Addition and Multiplication Facts

	Addition	Multiplication
	$0 + 0 = 0$	$0 * 0 = 0$
	$0 + 1 = 1$	$0 * 1 = 0$
	$0 + 2 = 2$	$0 * 2 = 0$
	.	.
	.	.
	.	.
	$4 + 4 = 3$	$4 * 4 = 1$

5. Write a program to print out all the values of INT (B/A) as A and B take on the values 1 through 10.

6. Write a program for printing out the square roots of all the integers between any two integers given by the user of the program providing the integers are not more than 25 apart.

7. Using the results of Homework Exercise 2 of this section, rewrite the program for finding divisors of a given positive integer making use of the INT function.

Exercises 4, 5, 6, and 7 of the preceding homework exercises involve "looping." Looping is repetition of a procedure a number of times and is the real power of a computer. The ability to perform loops is one of the differences between a computer and a calculator.

Up to this point looping has been handled in a primitive way. The BASIC language has a pair of statements which are specifically designed to set up loops in a simple manner.

Suppose, for the purposes of discussion, you wish to print out a table of squares and square roots for the integers from 1 to 100. A program to do this using only statements we have already learned is given below.

Program 1

```
10 LET I=0
20 LET I=I+1
30 LET X=I**2
40 LET Y=SQR(I)
50 PRINT I,X,Y
60 IF I=100 THEN 80
70 GØ TØ 20
80 END
```

The BASIC statements we now introduce are the FØR, TØ, STEP, and NEXT statements. They are always used together to form loops. Their use for the previous program is illustrated below.

Program 2

```
10 FØR I=1 TØ 100 STEP 1
20 LET X=I**2
30 LET Y=SQR(I)
40 PRINT I,X,Y
50 NEXT I
60 END
```

The output for the two previous programs is identical. However, using the FØR, TØ, STEP, and NEXT statements shortened the program by two statements. On larger and more complex programs, the number of statements saved could be much greater.

When executing the second program, the computer sets I = 1 and proceeds through the sequence of statements until it reaches the "NEXT I" statement. The value of I is increased by 1 and the statements between the FØR and NEXT statements are executed again. When I = 100, the statements within the loop are executed one more time and then the program stops. The general form of the FØR-TØ-NEXT loop is:

```
FØR X=Y TØ Z STEP S
    .
    .
    .
NEXT X
```

where X, Y, Z, and S are any variables, expressions, or (with the exception of X) constants which are integers. The first execution of the loop is with X = Y and each time through the loop (i.e. each time the program reaches the NEXT X statement) S is added to X. When X reaches a value outside the specified range (Y TO Z), the program proceeds beyond the loop.

A schematic picture to illustrate the flow within a loop is given in Figure 4.21.

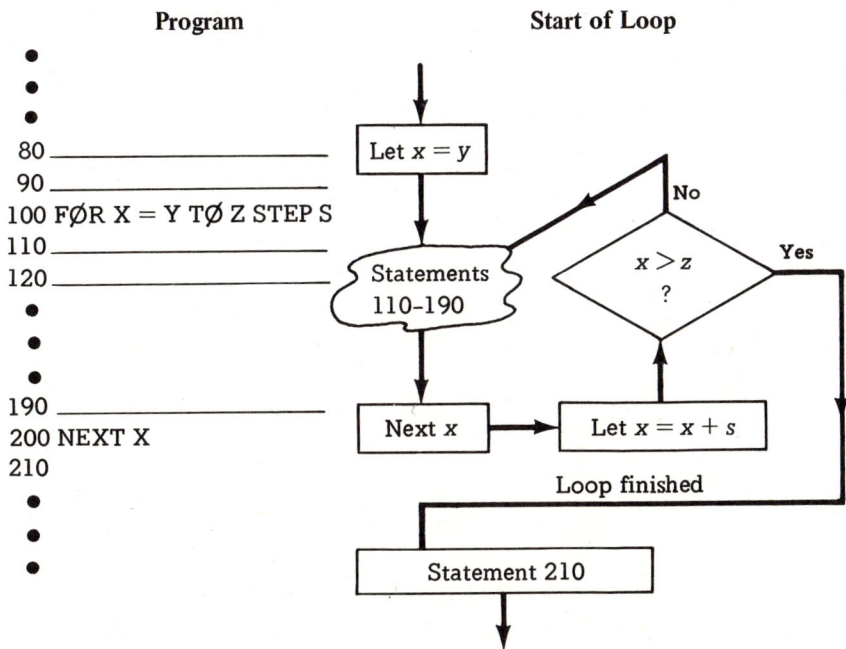

FIGURE 4.21

Note: There must be exactly one NEXT statement for each FØR statement in a program.

In flowcharting, FØR-TØ-NEXT loops can be represented many ways. For the sake of uniformity, represent loops as shown in Figure 4.22.

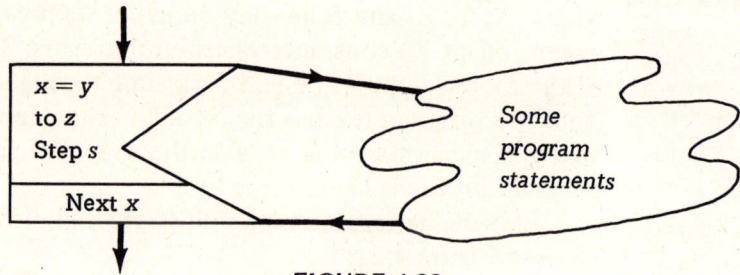

FIGURE 4.22

Using this convention, the flowchart for the previous program would look like Figure 4.23.

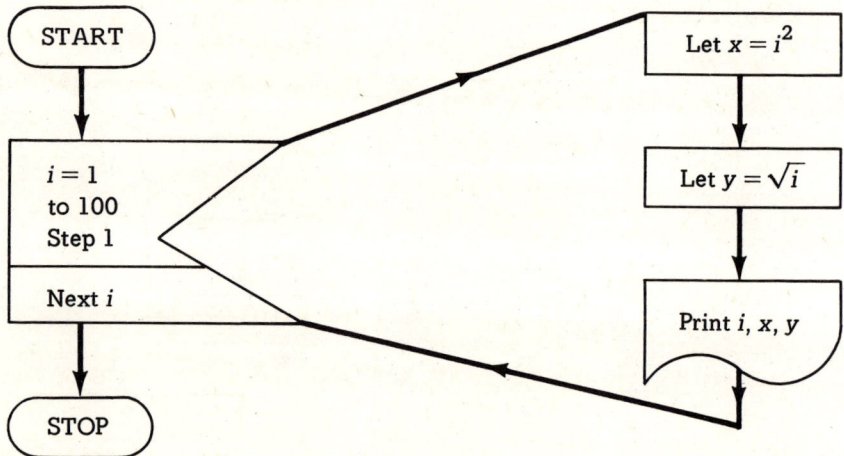

FIGURE 4.23

HOMEWORK EXERCISES

8. Redo the flowcharts for Exercises 4, 5, 6, and 7 of the previous homework exercises using the notational convention as shown in Figure 4.23.

If you wanted to print out only the squares and square roots of the odd integers between 1 and 100, you would simply change statement 10 of the previous program to read:

```
10 FØR I=1 TØ 100 STEP 2
```

1 would be the value of I the first time through the loop (statements 20, 30, 40, and 50), 3 would be the value of I the second time, and 99 would be the value of I the last time the loop is executed since 99 + 2 = 101, which is outside the specified range, 1 to 100.

If the STEP you wish is 1, you may omit the step designation completely and it is assumed to be 1 by the computer. For example, in Program 2 of this section, statement 10 could be abbreviated to:

```
10 FOR I=1 TØ 100
```

HOMEWORK EXERCISES

9. Rewrite the program for finding the divisors of a positive integer making use of FØR-TØ-NEXT loops. (**Hint**: loops can be "nested" completely inside other loops.)

10. Redo your programs for Homework Exercises 4, 5, 6, and 7 in Section 4.12 to make use of FØR-TØ-NEXT loops.

11. Flowchart the procedure and write a program for finding the prime numbers from 2 to K, where K is any positive integer greater than 1.

12. Flowchart the procedure and write a program for finding and printing out the prime factorization of any positive integer greater than 2.

13. Flowchart the procedures and write programs for finding the gcd and lcm for a given pair of positive integers.

★4.13   A LITTLE BIT MORE LOGIC

Sometimes, it is very difficult to prove directly that a statement is true. No doubt, you have experienced this difficulty at one time or another! Often, in these cases, however, it is easy to show that the opposite of the statement is false. This method is called an indirect proof or a *proof by contradiction.* In calculus, for example, there are some theorems that take over 25 pages to prove directly, yet take only a page to prove indirectly. The steps involved in a proof by contradiction are:

1. Assume that the opposite of the statement you are proving is true.

2. Use this assumption to arrive at a statement that contradicts either your assumption or a previously proved result.

3. Conclude that because of this contradiction, your assumption must be false and hence, the theorem is true.

This may all sound a little mystical; hopefully, the following example will provide some clarification.

   Example:

   PROVE: If $b \neq c$, then $a + b \neq a + c$ for a, b, $c \in W$.

   PROOF: Assume $b \neq c$ and $a + b = a + c$. Then by the cancellation law for whole number addition $a + b = a + c \Rightarrow b = c$ $\times$ *contradiction*!

We assumed $a + b = a + c$. Hence our assumption is false, and our theorem is true.

   Q.E.D.

Note: The symbol $\times$ is supposed to represent crossed swords and customarily designates the point of contradiction. The Q.E.D. comes from the Latin *quod erat demonstrandum,* which translates into which was to be demonstrated. It is conventional to use these letters to signify the end of a proof.

PROBLEM 1   Try proving this one by contradiction: If $2 \nmid n$, then $2 \mid n + 1$, for $n \in W$.

HOMEWORK EXERCISES

Prove the following statements by contradiction:

1. If $n = 2x + 1, x \in W$, then $2 \nmid n$.

2. If a gcd $b = 1$ and p is a prime, then p does not divide both a and b.

This proof of the theorem, "There is no largest prime number," provides another example of a proof by contradiction.

THEOREM: There is no largest prime number.

PROOF: Suppose that there is a largest prime number.

This is the same as saying that there is a finite number of primes. Represent these primes by $p_1, p_2, p_3, \ldots, p_k$. Now let $n = (p_1 \cdot p_2 \cdot p_3 \cdot \cdots \cdot p_k) + 1$. Then n is composite because n is greater than p_k which is the largest prime number. Since n is composite, it has prime factors. Let q be a prime factor of n. If q is prime, it must equal one of the p's and so $q \mid (p_1 \cdot p_2 \cdot p_3 \cdot \cdots \cdot p_k)$. If $q \mid n$ and $q \mid (p_1 \cdot p_2 \cdot p_3 \cdot \cdots \cdot p_k)$ then by the theorem you proved in the homework, $q \mid [n - (p_1 \cdot p_2 \cdot p_3 \cdot \cdots \cdot p_k)]$. Since $n - (p_1 \cdot p_2 \cdot p_3 \cdot \cdots \cdot p_k) = 1$, we have $q \mid 1$ ✕ Contradiction! This is impossible. We assumed that q is prime and hence is greater than one. Thus by assuming that there is a largest prime number, we have reached a contradiction of a known fact, namely that no number greater than one can divide one. Hence, our assumption that there is a largest prime number is false, which means that there is no largest prime number. Q.E.D.

HOMEWORK EXERCISES

3. Create a set of directions that could result in the following array being used as a sieve for primes.

TABLE 4.7

1	2	3	4	5	6
7	8	9	10	11	12
13	14	15	16	17	18
·	·	·	·	·	·
·	·	·	·	·	·
·	·	·	·	·	·

4. To find the gcd of numbers *a* and *b*, form a table with all factors of *a* and *b* listed across the top margin of the table and the numbers *a* and *b* listed in the left margin. Record a 1 in the body of the table for each factor of *a*, similarly for *b*. The gcd is the product of those factors having a 1 recorded for *a* and *b*. **Note:** Powers of one prime must be written out, i.e., $2^2 = 2 \cdot 2$.

Example: Find the gcd of 60 and 10.

TABLE 4.8

	2	2	3	5
60	1	1	1	1
10	1			1
gcd	1			1

$\Rightarrow$ gcd = $2 \times 5 = 10$

5. Try developing a procedure for lcm.

6. Decide whether lcm or gcd would be helpful in solving the following:

   a. $\dfrac{1}{8} + \dfrac{1}{32}$

   b. Finding a simpler expression for $\dfrac{8}{32}$.

★7. If it is possible to construct a set of consecutive nonprime whole numbers of length *k*, determine a rule for constructing such a set of *k* such integers, using the examples in Table 4.9 as a guide.

TABLE 4.9

n	Required Set	Intermediate Results
1	$\{4\}$	2 + 2
2	$\{8, 9\}$	6 + 2 6 + 3
3	$\{26, 27, 28\}$	24 + 2 24 + 4 24 + 3
4	$\{122, 123, 124, 125\}$	120 + 2 120 + 4 120 + 3 120 + 5
. . .		
k		

4.14 DIVISIBILITY TESTS

Remember when you were doing factor trees for large numbers? Didn't it seem that there should be a better method for determining what numbers divided those big numbers other than trial and error? The purpose of this section is to develop rules for divisibility by certain numbers.

PROBLEM 1   **a.** The following numbers are all divisible by 2: 18, 26, 42, 50, 174, 36, 212, 6. Complete the following rule for divisibility by 2: A number is divisible by 2 if and only if _____ .

b. That should have been easy; the next two rules should be more challenging. The following numbers are divisible by 4: 28; 4020; 15,976; 573,048; 728; 41,916; 312. Complete this rule: A number is divisible by 4 if and only if _____ .

c. The following numbers are all divisible by 3: 9; 111,111; 417; 456,789; 93,000,000. Complete the rule: A number is divisible by 3 if and only if _____ .

d. Did you have a tough time with the second or third rule? To make these two rules, as well as others, easier to prove, you are going to express numbers in a way that you may not have seen before. This new way of writing numbers is called *expanded notation* or *polynomial form*.

When you write the number 212 does each 2 mean the same thing? _____ Of course not! The two on the left means 2 times _____, and the 2 on the right means 2 times _____. The one, of course, means 1 times _____.

You could therefore write 212 as $2 \cdot 100 + 1 \cdot 10 + 2$, which is the expanded form of 212.

HOMEWORK EXERCISE

1. Write each of the following in expanded form.

    37 _____

    370 _____

    2942 _____

    93,000,000 _____

    abcd _____ (In this context *abcd* represents a four digit numeral, not the product of *a, b, c,* and *d.*)

PROBLEM 2   a. Notice that each of the places can be written as 10 to some power, e.g., $100 = 10^2$. Write each number below in the form 10^n where n is a natural number.

   $1000 =$ $10,000 =$ $100,000 =$ $10 =$

   b. Now keeping in mind the theorems, "if $a \mid b$ and $a \mid c$ then $a \mid (b + c)$," and "if $a \mid b$ then $a \mid bc$," you are ready to further investigate divisibility rules.

   In your first divisibility rule, you said that 174 was divisible by 2. Write 174 in expanded notation, and show why it is divisible by 2.

   You have just proven that $2 \mid 174$.

   c. Prove, in a similar fashion, that $2 \mid ab8$ where a and b can be any two integers between 0 and 9. **Note:** $ab8$ is one number, not a product of three numbers.

   In general, you can say $2 \mid abcd$ if and only if _____, or that any number is divisible by 2 if and only if _____.

172

In your very last statement, you assumed that if the number has a digit in the ten-thousands place it didn't matter. It doesn't, really, but only because $10,000 = 1000 \cdot 10$ and since $2 \mid 10$ then $2 \mid 1000 \cdot 10$. In general, $10^n = 10^{n-1} \cdot 10$ so $2 \mid 10^n$, provided $n \geqslant 1$.

d. Looking at *ab,cde* in expanded form, determine which digits need to be restricted for *ab,cde* to be divided by 4.

Complete the following rule: *ab,cde* is divisible by 4 if and only if _____. In a proof somewhat similar to your proof for divisibility by 2, prove the rule you have just stated.

Will your rule apply to a number with 7 digits? _____14 digits? _____Why, or why not?

e. Divisibility by 3 has a proof using a different technique than the proofs for 2 and 4. Prove that 261 is divisible by 3. **Hint:** In expanded notation, let $2 \cdot 100$ be replaced by $2(99 + 1)$.

Repeat your proof for divisibility by 3 on the number *abc,def* by writing it in expanded notation and using the previous hint.

In general $3 \mid abc,def$ if and only if _____. Can this rule be applied to a 7 digit number? _____ A 4 digit number? _____. Prove the rule for divisibility by 3 for *abcd*.

f. It's easy to check for divisibility by 5. If you don't know this rule, start counting by fives and it should come to you. A number is divisible by 5 if and only if _____. Prove your rule for 5 for the 4 digit number, *abcd*.

g. Since $8 = 4 \cdot 2$, the divisibility rule for 8 should be similar to the rules for 2 and 4. It is. Investigate *abc,def* in expanded notation and state a divisibility rule for 8. _____. Prove your rule for *abc,def*.

Is your rule valid for an 8 digit number? _____an 11 digit number? _____ .

★PROBLEM 3 a. Seven, 11, and 13 are very interesting primes. Would you believe that any number of the form *abc,abc* (i.e. 123,123) is divisible by 7, 11 and 13? Try a few and see if it works. Try to prove that *abc,abc* is divisible by 7, 11, and 13. **Hint:** $7 \cdot 11 \cdot 13 = 1,001$.

b. A beginning mind reader asks a person to think of a number from 1 to 999, multiply it by 143 and state the last digits of the product. Once this is done, the mind reader promptly states the original number. Can our friend really read minds? _____What is the gimmick? **Hint:** Remember the previous paragraph?

HOMEWORK EXERCISES

2. State and prove the divisibility rule for 9 on the number *ab,cde*. **Hint:** 9 = 3 · 3 so look at proof of divisibility rule for 3.

3. State and prove the divisibility rule for 10 on *abc,def*.

4. State a divisibility rule for 6.

5. Indicate which of the following are divisible by 2, 3, 4, 5, 8, 9, 10. Each number may have more than one answer.
   a. 261 is divisible by _____
   b. 240 is divisible by _____
   c. 93,000,000 is divisible by _____
   d. 816 is divisible by _____
   e. 51 is divisible by _____

6. Fill in the missing digits so that the integer is divisible by the indicated number.
   a. 2311_____by 3
   b. 2_____178 by 9
   c. 510, 35_____by 10
   d. 7_____ , 10 _____ by 3 and 4
   e. 95,_____16 by 8

7. Answer each of the following as true or false.
   a. _____ If *n* is divisible by 4, it is divisible by 2.
   b. _____ If *n* is divisible by 3, it is divisible by 9.
   c. _____ If *n* is divisible by 8, it is divisible by 4.
   d. _____ If *n* is divisible by 5, it is divisible by 10.
   e. _____ If *n* is divisible by 4 and 6, it is divisible by 8.
   f. _____ If *n* is divisible by 4 and 6, it is divisible by 12.
   g. _____ If *n* is divisible by 8 and 10, it is divisible by 40.
   h. _____ If *n* is divisible by 8 and 10, it is divisible by 16.
   i. _____ If *n* is divisible by 8, it is divisible by 2 and 4.
   j. _____ If *n* is divisible by 2, it is divisible by 6.

★8. Suppose that in going through your grandfather's old bills you come across one from the town butcher. It seems grandpa bought 72 turkeys as prizes for a church function about 75 years ago. The total at the bottom of the bill is faded and all that can be read is, $ _____67.9_____ where the blanks stand for digits that can no longer be read. Find the missing digits.

4.15 THE DIVISION AND EUCLIDEAN ALGORITHMS

Throughout your mathematical career, beginning with arithmetic in elementary school, you may have had problems with division. One problem is how to find the right quotient. The two division algorithms illustrated may help clarify this.

Examine the following division problem:

$$
\begin{array}{r}
242 \quad \text{R}30 \\
230\ \overline{)\ 55690} \\
46000 \\
\hline
9690 \\
9200 \\
\hline
490 \\
460 \\
\hline
30
\end{array}
\qquad
\begin{array}{l}
200 \times 230 = 46000 \\[10pt]
40 \times 230 = 9200 \\[10pt]
2 \times 230 = 460
\end{array}
$$

The idea of finding a largest quotient and smallest remainder is made mathematically precise by the following statement.

> *DIVISION ALGORITHM:* For every $a, b \in Z$, $b > 0$, there exists $q, r \in Z$ such that $a = q \cdot b + r$ where $0 \leqslant r < b$. (q can be thought of as the quotient and r as the remainder.)

The long division process consists of repeated application of the division algorithm. The division is often shortened to this form:

$$
\begin{array}{r}
242 \quad \text{R}30 \\
230\ \overline{)\ 55690} \\
460 \\
\hline
969 \\
920 \\
\hline
490 \\
460 \\
\hline
30
\end{array}
$$

This process is not an easy one to understand. It subtly assumes that you are fully aware of the place value of each digit in 55690. This is the reason that we seemingly start out by dividing

556 by 230 and getting 2. We are really dividing 55600 by 230 and getting 200. Now this is not too difficult for us, but it is often far too much to assume that elementary school children can grasp this, hence the problems with teaching division. Another problem is that one must continually find the largest q and the smallest r as specified in the division algorithm.

Both of these problems can be alleviated by dividing by a slightly different process. We use our original example as an illustration:

$$
\begin{array}{rl}
230 \overline{)\ 55690} & 100 \\
23000 \\
\overline{32690} & 100 \\
23000 \\
\overline{9690} & 40 \\
9200 \\
\overline{490} & 2 \\
460 \\
\overline{30} \quad \overline{242} & \text{R } 30
\end{array}
$$

The train of thought here is the following: find a number which when multiplied by 230 is less than 55600, in this case 100. Subtract the product from 55690. Now find a number which when multiplied by 230 is less than 32690; again, 100 works. Repeat this procedure until you can't find an integer to multiply times 230 that will result in their product being less than the last difference. This last difference is the remainder (in this case, 30). The sum of the various numbers you have found (the column on the right) is the quotient. **Note:** You may have chosen different numbers from the ones illustrated.

The advantages of this process are that it illustrates the idea of place value and that it does not successively require the largest q.

PROBLEM 1   Try the following using this new process:
a. $63 \overline{)\ 4973}$
b. $105 \overline{)\ 11365}$
c. $327 \overline{)\ 567943}$

176

HOMEWORK EXERCISES

1. Determine how many applications of the division algorithm are necessary to perform each of the following (using the first process). Find the q's and r's for each application.
   a. $65 \overline{)335}$
   b. $17 \overline{)289}$
   c. $43 \overline{)23693}$

2. Using the second process, perform each of the following:
   a. $73 \overline{)23942}$
   b. $18 \overline{)3974}$
   c. $67 \overline{)82931}$

Before letting the division algorithm go, one last application will be shown. Euclid knew about the division algorithm and ingeniously saw in it a method for finding the gcd of 2 numbers. He developed the process and called it the *Euclidean Algorithm.* The process works especially well for large numbers. Here is an example:

Find 2431 gcd 323

$$2431 = 7 \cdot 323 + 170$$
$$323 = 1 \cdot 170 + 153$$
$$170 = 1 \cdot 153 + \boxed{17} \leftarrow 2431 \text{ gcd } 323$$
$$153 = 9 \cdot 17 + 0$$

PROBLEM 2 a. Check the results obtained with your own gcd algorithm. Does it seem to work?
   b. Imitate the procedure and find the gcd of 999 and 711. Here is a start:

$$999 = 1 \cdot 711 + 288$$

What did you get as the greatest common divisor? If you got 9, you're doing fine! This process is completely general and will work for any two whole numbers. Starting with the last equation and working your way up, describe how you can verify that your answer is a common divisor. **(Hint:** If $a \mid (b - c)$ and $a \mid c$, then $a \mid b$). Explain why it must be the greatest common divisor.

   c. Use the Euclidean Algorithm to find a gcd b in each of the following. Check your answer with your own algorithm where feasible.

     1. $a = 75$
          $b = 30$

     2. $a = 125$
          $b = 626$

     3. $a = 1456$
          $b = 3536$

HOMEWORK EXERCISE

Find a gcd b by any method.

3. $a = 65$
     $b = 26$

4. $a = 437$
     $b = 323$

5. $a = 1309$
     $b = 165$

S U M M A R Y

At the heart of Chapter 4 is the concept of relations and the way in which relations structure different sets. Relations are more fundamental than the binary operations examined in Chapters 2 and 3. Relations are one of the very simplest concepts providing for ways to relate two elements of a set. Some of the common relations—equal, less than, parallel, perpendicular, collinear, and factor—were discussed together with their properties. As might be expected for such a basic concept as relation, a wide variety of ways have been developed to express this idea such as directed graphs, sets of ordered pairs, Cartesian coordinate graphs, a listing, a criterion. The processes of mathematics were constantly utilized in working with all these ideas. In this chapter another deductive system was developed—this time in a finite geometry. The use of both inductive and deductive processes is a familiar pattern by now.

Sets Structured by Mappings

5

5.1 · SYMMETRY AND MOTION

The most appealing patterns in natural objects, geometric figures, music and art, or collections of numbers very often embody some type of symmetry or balance. For example, each of the designs in Figure 5.1 has what scientists call bilateral symmetry or symmetry with respect to a line.

In this section you will be asked to identify symmetry in various geometric shapes, to develop a procedure for checking symmetry, and to experiment with methods for creating symmetric figures.

PROBLEM 1 Examine the designs in Figure 5.2. Using your intuitive ideas of symmetry, decide which of the figures have one or more types of symmetry and which have none.

PROBLEM 2 Thirteen of the designs have symmetry. After re-checking your decisions, formulate, in your own words, definitions of each type of symmetry and a procedure for actually testing for the presence of symmetry.

One simple general test for symmetry makes use of tracing and repositioning of the design:

A tone row and its retrograde
12-tone mustic

```
        1
       1 1
      1 2 1
     1 3 3 1
    1 4 6 4 1
  1 5 10 10 5 1
• • • • • • • • •
```

Pascal's triangle

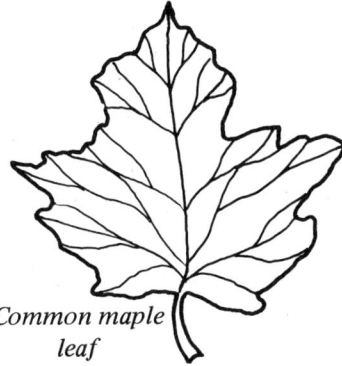

Common maple
leaf

FIGURE 5.1

1. Trace the design in question on tracing paper.
2. Try to shift the tracing to a new position that coincides with the original design.
3. If you find such a shift, the figure is symmetric; if not, the figure has no symmetry.

Of course, any tracing coincides with the original design if you simply lift the tracing, move it around, and place it back on the original in the same position; so the key is to find a different position for the tracing paper that leaves the original design covered by the tracing of that design. You may turn, slide, or flip the paper to obtain the new position; but cutting, tearing, or wrinkling are not permitted. You may use any *rigid motion* of the tracing paper to find a new position. Remember you used these rigid motions in connection with some geometric figures in Section 3.5. The definition you used follows:

DEFINITION: Any figure for which a rigid motion, other than the identity, exists is said to be *symmetric*. Each differ-

181

ent rigid motion, other than the identity, shows a *symmetry* of the figure.

PROBLEM 3   One kind of symmetry is *bilateral* or *line symmetry,* which is often also called *mirror symmetry.*
a. Which of the designs in Figure 5.2 have this type of symmetry?

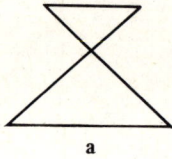

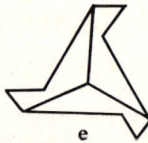

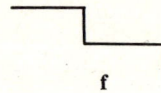

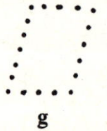

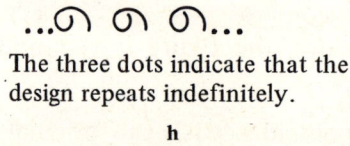

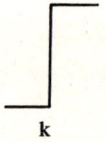

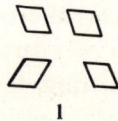

...ꓜ ꓜ ꓜ...

The three dots indicate that the
design repeats indefinitely.

h

FIGURE 5.2

b. Use a small pocket mirror to confirm your decision, and to show the appropriateness of the term mirror symmetry.

c. What type of rigid motion of tracing paper tests for line symmetry? Demonstrate your technique on the designs in Figure 5.2.

d. The flips of an equilateral triangle in Section 3.5 are examples of rigid motions that show line symmetry. How many different rigid motions of an equilateral triangle show line symmetry? Identify the lines of symmetry in an equilateral triangle.

PROBLEM 4 A second important type of symmetry is radial or rotational symmetry, which is often also called point symmetry.

a. Which designs in Figure 5.2 have this type of symmetry?

b. What rigid motion of the tracing paper confirms rotational symmetry? This also shows the appropriateness of the term *point* symmetry.

c. The turns of an equilateral triangle in Section 3.5 are examples of rigid motions that show rotational symmetry. How many different rigid motions of an equilateral triangle showed rotational symmetry? Locate the center of rotational symmetry in an equilateral triangle.

PROBLEM 5 If a slide of the tracing paper positions the tracing design exactly over the original design, the design has translational or slide symmetry.

a. Which designs in Figure 5.2 have slide symmetry?

b. How are the three dots essential to slide symmetry?

PROBLEM 6 Many familiar figures have several symmetries; they even have different types of symmetries. Consider Figure 5.3.

FIGURE 5.3

a. Describe the rigid motions that show line symmetries of the star. How many are there? Locate the lines of symmetry.

b. Describe the rigid motions that show rotational symmetries of the star. How many are there? Locate the center of the rotational symmetry.

c. Describe the rigid motions that show translational symmetries of the star. How many are there? Identify the direction and distance of the translation.

PROBLEM 7   Repeat the same process used in Problem 6 for describing all the symmetries of a dot pattern. You might want to use tracing paper as in Problem 5.

FIGURE 5.4

★PROBLEM 8   Repeat the same process used in Problem 6 for describing all the symmetries of the following grid which covers the entire plane.

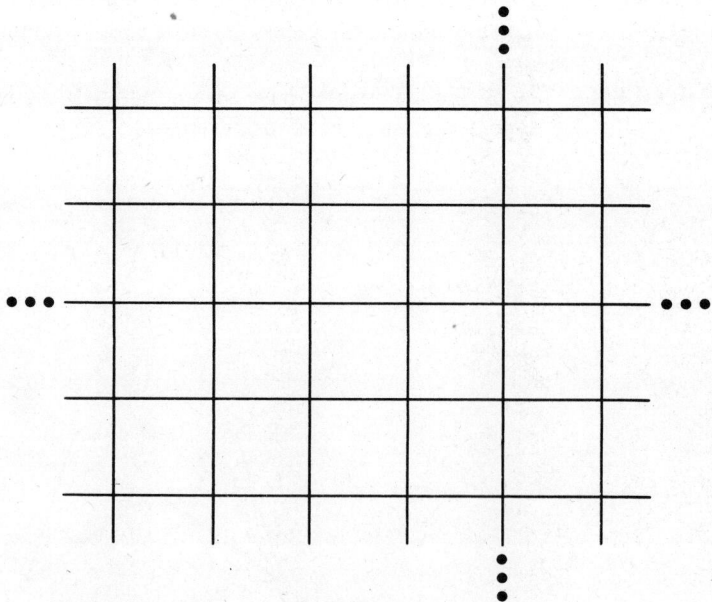

FIGURE 5.5

184

★PROBLEM 9   Some of the figures you have examined have both rotational and line symmetry. Formulate a conjecture which relates the center of rotational symmetry and the lines of symmetry.

HOMEWORK EXERCISES

1. Describe all the symmetries of a six-pointed star.

FIGURE 5.6

2. Describe all the symmetries of a sine wave such as you might see in a pattern of ocean waves, on an oscilloscope, or fluttering across a dying television picture.

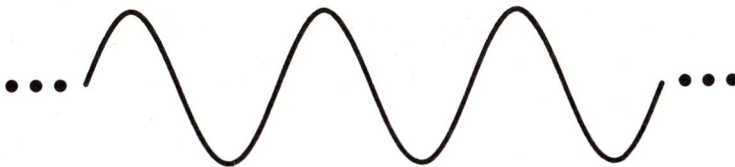

FIGURE 5.7

3. Draw a polygon that has exactly one symmetry. Remember the identity rigid motion does not define a symmetry by the definition used. Also, for now, a *polygon* is a simple, closed figure with line segments for edges.

185

★4. Draw a polygon that has exactly two symmetries.

★5. Draw a polygon that has exactly three symmetries.

★★6. What are the maximum and minimum number of symmetries that can occur in an *n*-sided polygon? Start your investigation by considering several simple cases such as triangles and quadrilaterals. When you detect a pattern to the results, try to generalize it.

LABORATORY PROJECTS

1. Obtain a set of *Mirror Cards* and use your knowledge of bilateral or line symmetry to solve the mirror placement problems in several sets. Then construct a pattern card and mirror cards for your own design. *Mirror Cards* are available from Educational Development Corporation, Inc., Newton, Massachusetts.

2. Construct the letters of the alphabet. Using mirrors and/or tracing paper, determine which letters have bilateral or line symmetry and which letters have rotational or point symmetry. The way you construct your letters will affect your answers.

Notice that, again, you are using many of the important thinking processes in mathematics many times in this section. You formulated generalizations, tested conjectures, and experimented with physical testing and manipulation. The work with symmetry is particularly rich for physical manipulation. Often a mathematical concept will become clear from some simple physical manipulation.

5.2 MAPPING

PROBLEM 1   Create a symmetric figure consisting of eight dots, using the indicated line as a mirror line. Draw an arrow from each of the given dots to its mirror image. The arrows are used to call attention to the link between the given dots and their images.

186

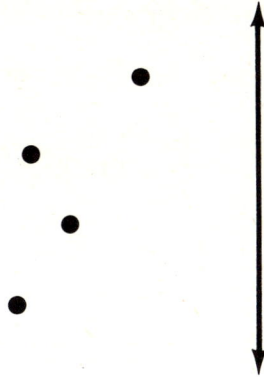

FIGURE 5.8

PROBLEM 2   Show the location for the given dots if each one is slid horizontally two inches to the right. Draw an arrow from each of the given dots to its new location. By analogy with the preceding example, the dots in the new location are said to be the images of the original dots.

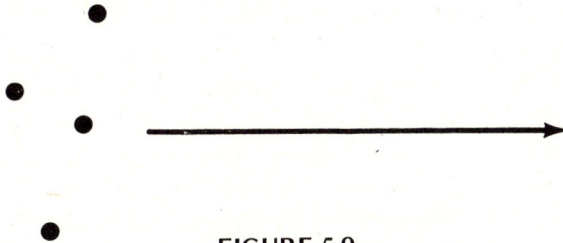

FIGURE 5.9

In Problem 1 and Problem 2, a set of elements is linked in a special way to another set of elements called the *image* set. The arrows call attention to the one-directional nature of this link, which goes *from* the original set *to* the image set. This one-directional pairing of elements is called a *mapping* or *function*.

DEFINITION: A mapping is a rule which assigns to each element of a set *A* a unique element of a set *B*.

Note: *Unique* means that one and only one element is assigned.

187

Following are some examples of mappings:

Example 1

Set $A = \left\{ {}^-3,\ {}^-1,\ 2,\ 4 \right\}$ and set $B = \left\{ 1,\ 4,\ 9,\ 16,\ 25 \right\}$. The rule for assigning images is: the image of an element of A is the square of that element. Figure 5.10 is an arrow diagram for mapping sets A and B.

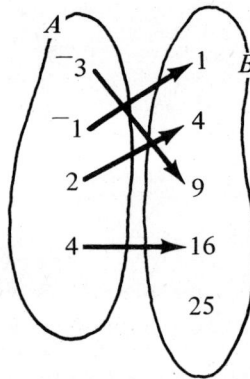

FIGURE 5.10

The rule for the mapping can be written algebraically as $x \rightarrow x^2$. Notice that in this mapping every element in set A has an image in set B, but not every element in set B is the image of some element in set A.

Example 2

In classes taught by the small group method, the class is separated into groups of four or five students, with each student working in a group. Consider set A to be all the students in the class and set B to be all the small groups. Then placing students in groups is an example of a mapping. The precise rule that determines the assignment may not be known, but the assignment can be completely specified by a diagram showing to which group each student belongs. Notice that in this mapping more than one element of set A has the same image in set B although no element of set A has more than one image in set B.

Example 3

Set A and set B in the definition do not have to be different sets. The rule which assigns to every integer its successor is a mapping of set Z to itself, which may be written $x \rightarrow x + 1$ for all $x \in Z$.

Set *A* of the definition is called the *domain* of the mapping. The set of images of elements of *A* is called the *range* of the mapping. This set of images is a subset of set *B*.

PROBLEM 3   Examine the arrow diagrams in Figure 5.11.

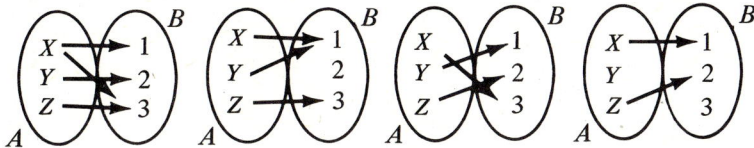

FIGURE 5.11

a. Decide which represent mappings. For those that do not, indicate why not.
b. Indicate the domain and range for each diagram that represents a mapping.

PROBLEM 4   Each of the following statements names two sets that are related. Next to each pair of sets write Yes if a mapping is represented, or No if a mapping is not represented.
a. A telephone directory listing of names of subscribers and their phone numbers. _____
b. A picture and an enlargement of it. _____
c. A listing of students' names and their social security numbers. _____
d. The teacher's roster listing students' names together with their scores on each of five tests. _____
e. A pairing of whole numbers and the largest prime factor of the number. _____
f. A pairing of the months of the year with the number of days in each month. _____
g. A code that changes each letter of the alphabet to a different letter of the alphabet according to a formula. _____
h. A city and a street map of the city. _____

You have probably already noticed that when sets *A* and *B* in the definition of a mapping are the same set, a mapping is nothing more than a special case of a relation. Consider the first name-last name relation used in connection with the directed

189

graph in Section 4.1. When defined on a set $K = \{$ Jane Smith, Stan West, Scott Moyer, Nancy Johnson, Bill Webb, Al Anderson, William Blake $\}$, the resulting ordered pairs form a simple relation. When defined on the set $T = \{$ Bill Mullins, Scott Anderson, Mary Johnson, Alice North, Jane Smith, William Blake, Nancy Webb $\}$, the resulting ordered pairs form a special kind of relation, a mapping.

PROBLEM 5   Fill in the arrows in the two arrow diagrams of Figure 5.12 using the rule: The first initial of the first name of . . . is the same as the first initial of the last name of . . .

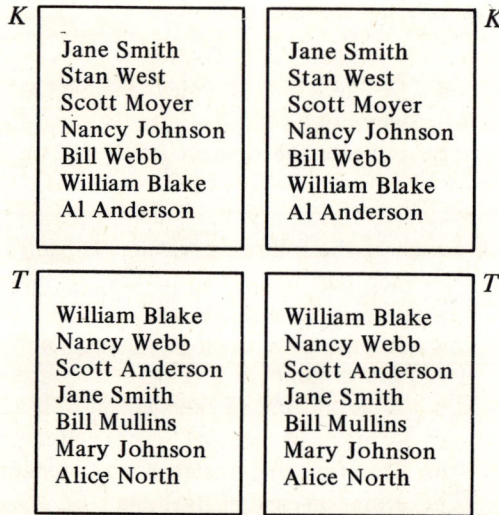

K ┌─────────────────┐   ┌─────────────────┐ K

Jane Smith	Jane Smith
Stan West	Stan West
Scott Moyer	Scott Moyer
Nancy Johnson	Nancy Johnson
Bill Webb	Bill Webb
William Blake	William Blake
Al Anderson	Al Anderson

T ┌─────────────────┐   ┌─────────────────┐ T

William Blake	William Blake
Nancy Webb	Nancy Webb
Scott Anderson	Scott Anderson
Jane Smith	Jane Smith
Bill Mullins	Bill Mullins
Mary Johnson	Mary Johnson
Alice North	Alice North

FIGURE 5.12

There really isn't much difference between an arrow diagram and a directed graph when set A and set B in the definition of a mapping are the same. The arrow diagram emphasizes the domain and the range sets separately; the directed graph emphasizes the relationship among the elements of one set. In both, the arrows indicate the ordered pairs in the mapping or relation.

PROBLEM 6   Consider the set of elements Z_5 as domain and the rule $x \rightarrow x + 1$ with addition defined in $(Z_5, +)$.

a. Draw an arrow diagram to show this mapping.
b. Name the range for this mapping.
c. Draw a directed graph for this mapping.
d. List all the ordered pairs formed by this mapping.

PROBLEM 7   Formulate a rule to determine when a relation is a mapping by simply looking at the directed graph of a relation.

PROBLEM 8   Test the rule you formulated on the following directed graphs and determine which relations are mappings.

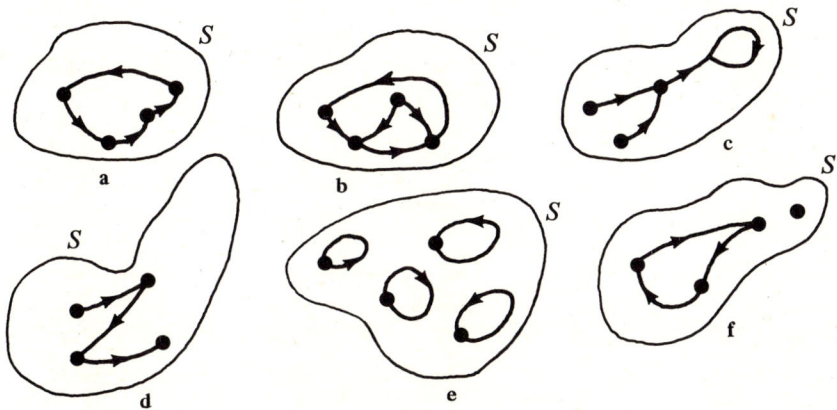

FIGURE 5.13

HOMEWORK EXERCISES

1. Create a symmetric figure consisting of eight dots, using the indicated dot as a center of rotation for a half turn (half of a full rotation). Draw an arrow from each of the original four dots to their images after the half turn.

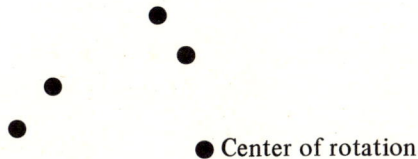

Center of rotation

FIGURE 5.14

2. Indicate which of the arrow diagrams in Figure 5.15 represent mappings. For those that do not represent mappings, indicate why not.

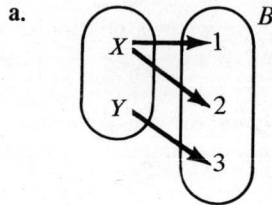

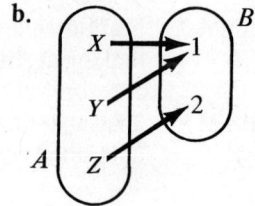

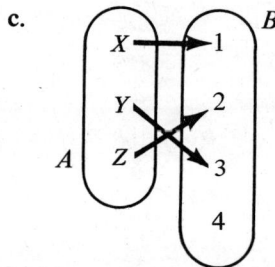

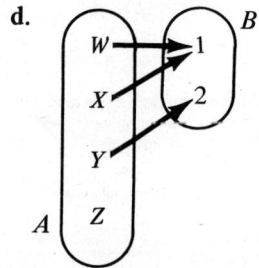

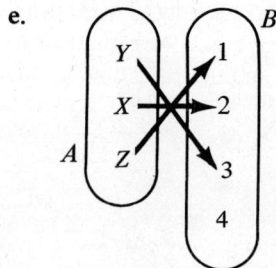

a.

b.

c.

d.

e.

FIGURE 5.15

3. For each of the following arrow diagrams that represent mappings, write
   the domain and range, and when possible give a rule for the mapping.

   a. $0 \to 0$

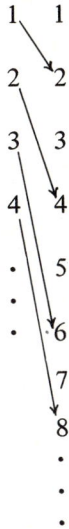

   b. $1 \to 1$

   $\dfrac{1}{2} \to 2$

   $\dfrac{1}{3} \to 3$

   $\dfrac{1}{4} \to 4$

   $\dfrac{1}{5} \to 5$

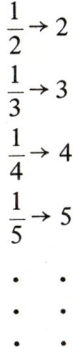

   c. $1 \to 1$

   $2 \to 3$

   $3 \to 6$

   $4 \to 10$

   $5 \to 15$

   $6 \to 21$

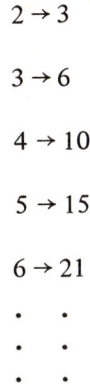

   d. $0 \to 0$

   $1 \to 1$

   $2 \to 2$

   $3 \to 3$

   $4 \to 4$

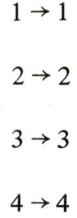

4. Give the domain and the range for each of the following four diagrams of
   mappings. If possible, give a rule for the mapping. **Note:** In Figure 5.18,
   zero is not considered part of the domain.

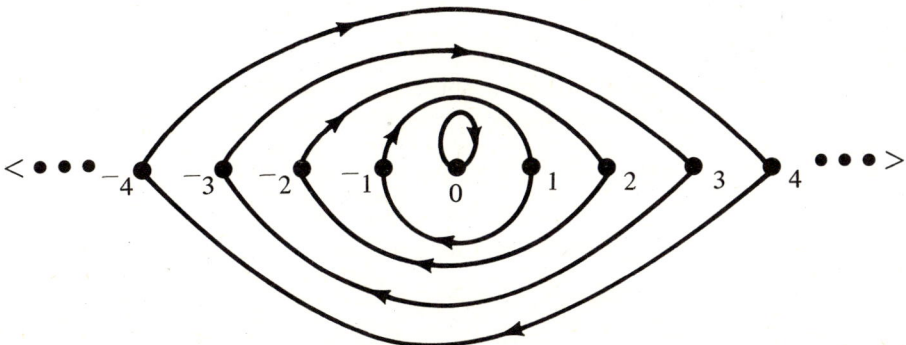

FIGURE 5.16

193

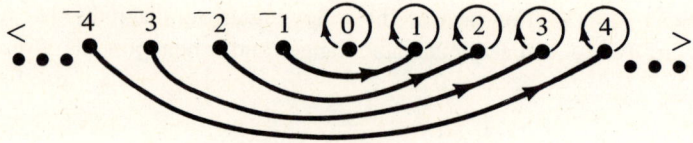

FIGURE 5.17

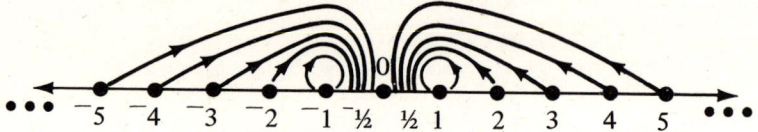

FIGURE 5.18

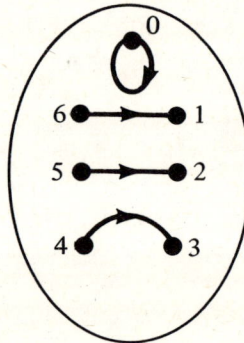

FIGURE 5.19

5. Decide if each of the following directed graphs is or is not a mapping. For those which are mappings, give domain and range. Then list all the ordered pairs of the mapping, and give a rule for the mapping.

a.   b.   c.   d.

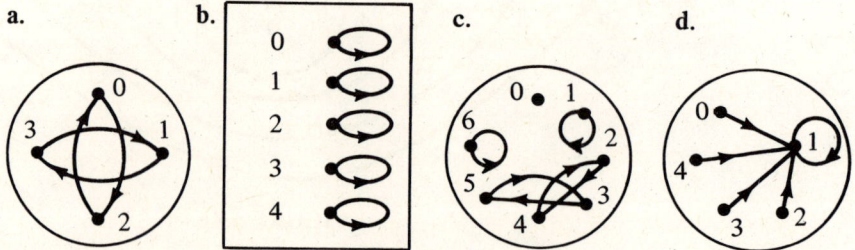

FIGURE 5.20

5.3   MAPPING GRAPHS AND APPLICATIONS

So far you've studied mappings that pair points and their images in symmetric figures or numbers and their images under various algebraic operations. The mapping assignments have been pictured by arrow diagrams, directed graphs, and lists of ordered pairs. The mapping family of special relations occurs in many other real life situations—often in two dimensional graphs or charts such as the following.

PROBLEM 1   The graph in Figure 5.21 indicates temperature fluctuation over a 24-hour period. It is a mapping from $T = \{$ 12 mid, 3 am, 6 am, 9 am, 12 noon, 3 pm, 6 pm, 9 pm $\}$ to Z.

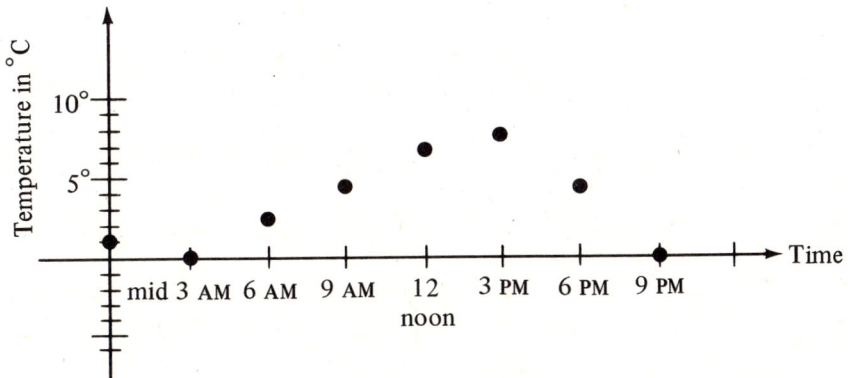

FIGURE 5.21

a. What was the 9 am temperature? The 6 pm temperature?

b. When was the temperature 5°?

c. What were the high and low temperatures and when did they occur? What was the temperature range?

★d. What was the average temperature during the day (24-hour period)?

★e. Does the graph specify a mapping from Z (domain) to T?

PROBLEM 2   Analyze the following mapping graph.

a. When did the market reach its peak during the given time?

b. What are the domain and range? **Hint:** The graph pictured is deceptive.

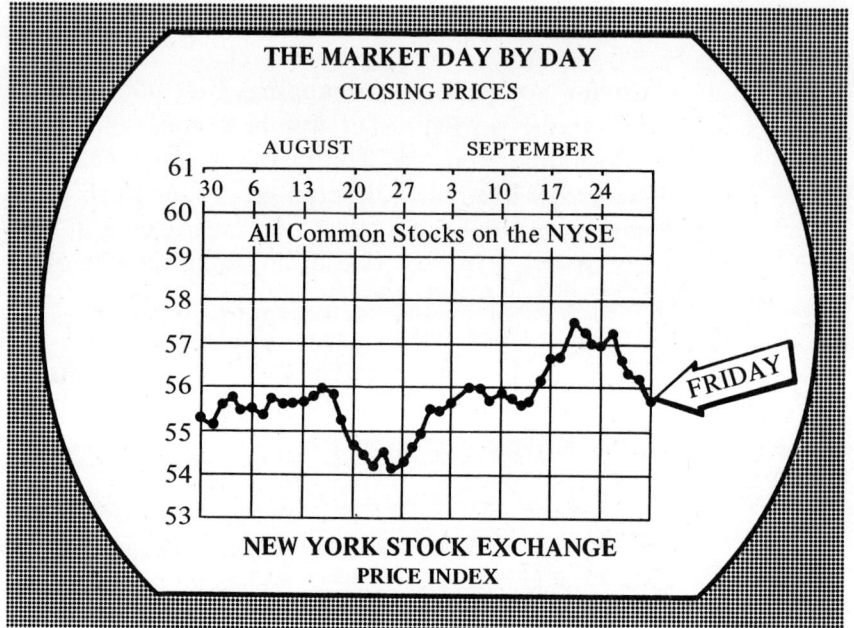

THE MARKET DAY BY DAY
CLOSING PRICES

AUGUST   SEPTEMBER

All Common Stocks on the NYSE

FRIDAY

NEW YORK STOCK EXCHANGE
PRICE INDEX

Courtesy of *The Washington Post*

FIGURE 5.22

    c. Discuss some reasons why the graph is drawn with the dots connected.

★**d.** Estimate the average price index value for the given time and then calculate the average to check your guess.

PROBLEM 3   The following graph is drawn from a report on population growth.

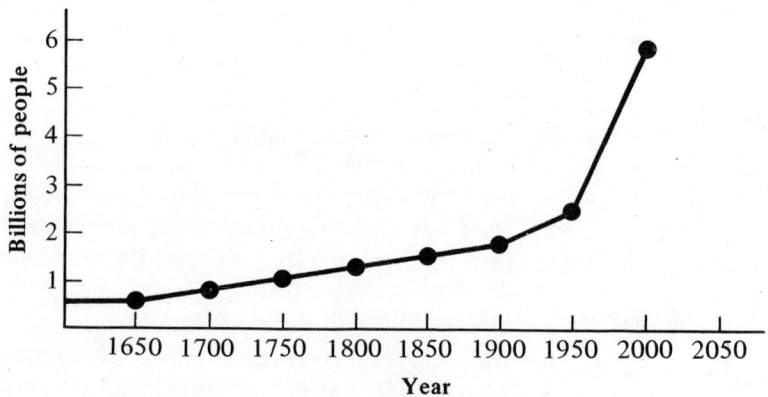

FIGURE 5.23

a. What are the domain and range of the mapping given by the graph? Your answer will depend on your interpretation of the graph.
b. When did the world population reach 2 billion? 3 billion?
c. What was the world population in 1700?
d. How accurate do you suspect the figures are? How might they have been obtained?

PROBLEM 4   Make a graph of the mapping from Z to Z with rule $x \to 2x$, using a coordinate grid like that pictured below. Start by making a table of some of the assignments and plot the corresponding points.

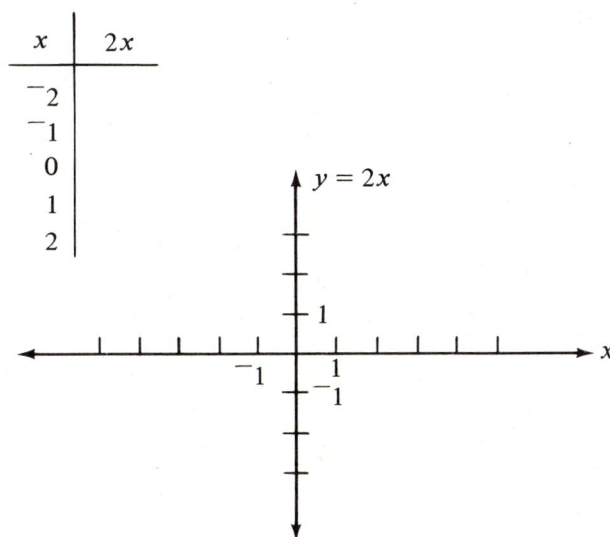

FIGURE 5.24

★PROBLEM 5   What is the relation between high school academic performance and first semester college grade point? Is there a mapping that assigns to each high school class ranking a GPA? If so, does the mapping have a simple algebraic rule? Study the data provided in Table 5.1 as a basis for answering the questions about GPA prediction. (The 60th percentile means 60% of class was below Student 1, etc.)

a. On a sheet of graph paper plot the data carefully.
b. Though the data will not lie in a straight line, you can draw a line that almost fits the trend of the numbers.
c. Using your best fit line as the mapping predictor, what GPA can be expected by a student whose high school class rank was 50th percentile? 10th percentile? How far off is your predicting line for student 1? for student 5?

197

TABLE 5.1

Student	H.S. Class Rank (percentile)	GPA First Semester
1	60	1.2
2	70	2.1
3	55	2.4
4	25	2.1
5	90	3.5
6	15	1.8
7	40	2.0
8	80	2.3
9	95	2.5
10	30	2.3

As the examples in Problems 1 to 5 show, mapping assignments and graphs occur in a wide variety of quantitative situations in which each element of a domain is paired with a single element of some suitable range.

HOMEWORK EXERCISES

1. For the information in Table 5.2, identify the mappings and determine the domain and range of each.

TABLE 5.2   Characteristics of electronic calculators on sale.

Manufacturer	Original Price	Sale Price	No. of Digits	% Key	Constant	Memory	√ Key	Log and Trig Functions
TI-1250	12.95	9.65	8	X	X	X		
TI-1200	9.95	7.35	8	X	X			
Royal 81M	13.95	9.75	8	X	X	X		
TI-1270	14.95	11.50	8			X	X	
Canon 8M	19.95	12.90	8	X	X	X	X	
TI-30	24.95	18.85	8	X	X	X	X	X
Canon 8	16.95	9.90	8	X	X		X	
Rockwell 64RD	39.95	28.50	12		X	X	X	X

(cont.)

TABLE 5.2 (cont.)

Manufac-turer	Original Price	Sale Price	No. of Digits	% Key	Constant	Memory	√ Key	Log and Trig Functions
Royal 91S	24.95	14.75	8	X	X	X	X	
Rockwell 44RD	29.95	19.75	9		X	X	X	X
Novus 6020	34.95	22.75	8	X	X	X		
Sharp EL8009	39.95	32.50	8		X			
Sharp EL8116	19.95	15.80	8	X	X	X	X	

2. Determine the domain and the range for the graph of a mapping in Figure 5.25.

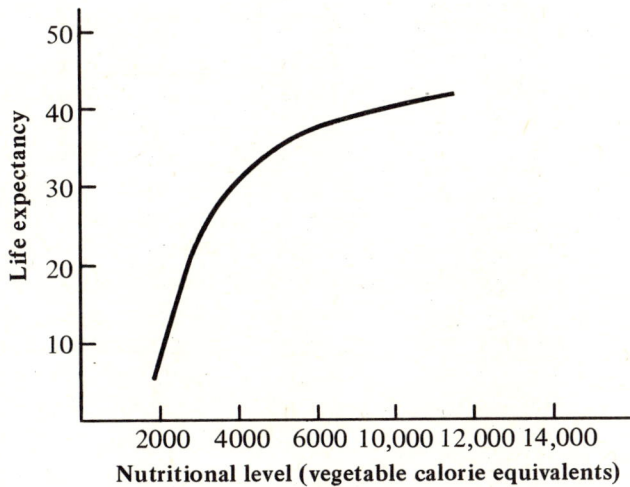

Based on data from 35 nations.

FIGURE 5.25

3. Make a coordinate grid graph of each of the following mappings for $^-5 \leqslant x \leqslant 5$, in Z.

  a. $x \rightarrow x$

199

   b. $x \rightarrow x - 1$
   c. $x \rightarrow x - 2$
   d. $x \rightarrow {}^{-}x$
   e. $x \rightarrow {}^{-}x - 1$
   f. $x \rightarrow {}^{-}2x$

4. Scout through newspapers, magazines, and books to locate at least 5 different examples of mappings graphed in various ways. Bring in the pictures or be prepared to explain them to the class.

5. Use the graph given in Figure 5.26 to complete the following list of number $\rightarrow$ image assignments:

   a. $0 \rightarrow$
   b. $1 \rightarrow$
   c. $4 \rightarrow$
   d. ${}^{-}7 \rightarrow$
   e. $\rightarrow {}^{-}2$
   f. $\rightarrow 5$
   g. $\rightarrow 0$
   h. $\rightarrow {}^{-}5$

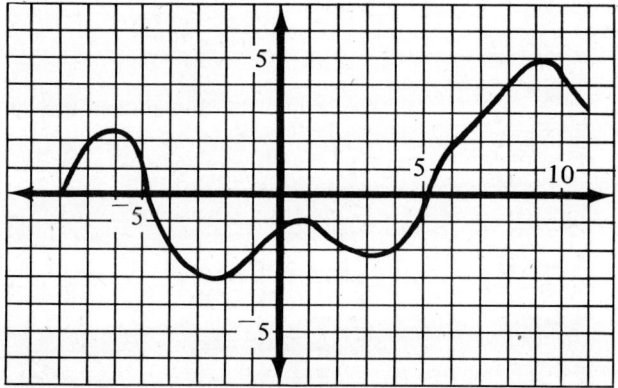

FIGURE 5.26

★6. Make coordinate grid graphs for the mappings $x \rightarrow x^2$, $x \rightarrow x^2 + 1$, and $x \rightarrow {}^{-}x^2$ for ${}^{-}3 \leqslant x \leqslant 3$, in Z.

5.4 INVERSES FOR MAPPINGS

As the United States moves slowly but steadily toward the metric system of measurement, you will note more and more reference to the Celsius scale of temperature measurement. Though you are quite familiar with the $0°$ freezing point and the $100°$ boiling point of water in the Celsius system, for intermediate temperatures many of you may be asking the question, "What's that in Fahrenheit?" for several years.

PROBLEM 1   The rule for converting Celsius to Fahrenheit temperature readings is of the form F = _____ · C + _____ . By trying several known matchups and others given on the scale at the right, determine the missing numbers in the mapping rule.

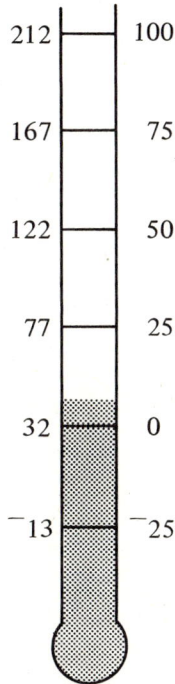

212	100
167	75
122	50
77	25
32	0
$^-$13	$^-$25

FIGURE 5.27

PROBLEM 2   The Celsius equivalent of certain Fahrenheit readings will soon be in common usage.
    a.   What is Celsius value of normal body temperature, 98.6°F?
    b.   What is Celsius value of room temperature, 68°F?
    c.   Find a rule for the Fahrenheit to Celsius mapping.

The two temperature converting mappings have a very useful and special relationship; each reverses the assignments of the other.

$$0 \rightarrow 32 \rightarrow 0$$

$$100 \rightarrow 212 \rightarrow 100$$

$$50 \rightarrow 122 \rightarrow 50$$

For this reason they are called *inverses* of each other.

PROBLEM 3   In what sense is the word *inverse* used similarly when reference is to mappings or to operations in groups?

PROBLEM 4   Not all mappings have inverses that are mappings. Study each of the following examples.

a.   The mapping is from Z_5 to Z_5 with a directed graph as below. What is the inverse? Is this inverse a mapping?

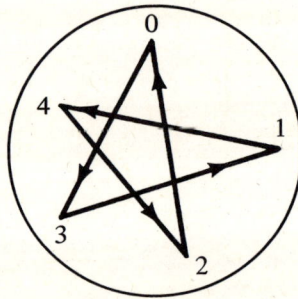

FIGURE 5.28

b.   The mapping is from Z to Z with the arrow diagram as shown in Figure 5.29. What is the inverse? Is the inverse a mapping?

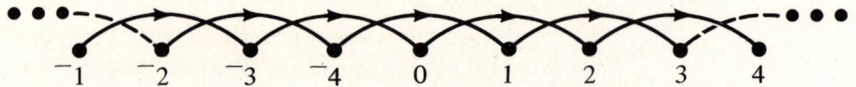

FIGURE 5.29

c.   The mapping is from Z to W as graphed in Figure 5.30. What is the inverse? Is the inverse a mapping?

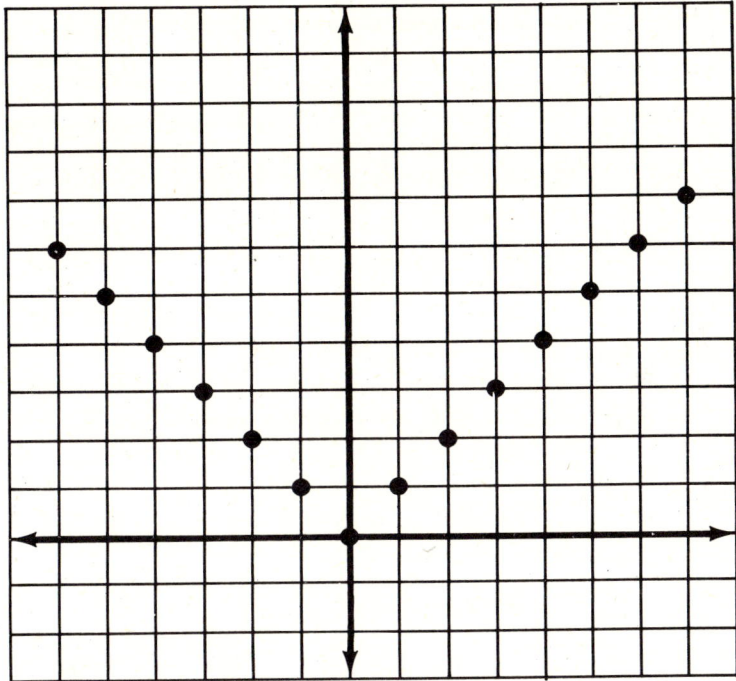

FIGURE 5.30

d. The mapping is from A to B with the arrow diagram as shown in Figure 5.31. What is the inverse? Is the inverse a mapping?

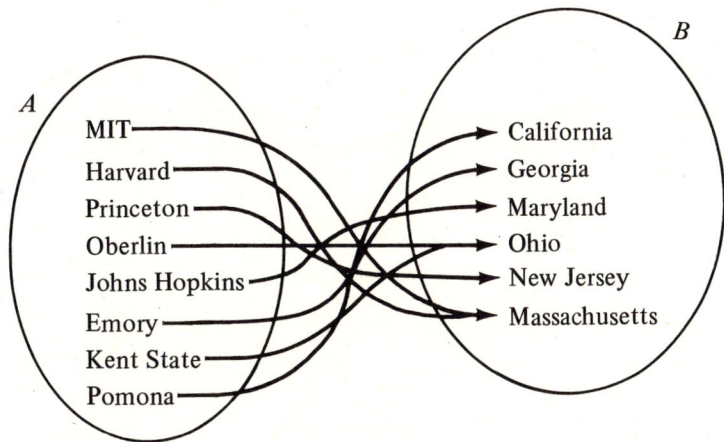

FIGURE 5.31

e. The mapping assigns each home in the U.S. a zip code for directing mail. What is the inverse? Is the inverse a mapping?

PROBLEM 5   What general characteristics distinguish the mappings that have inverses that are mappings from those that have inverses that are not mappings? You may want to refer to Problem 4 for examples to help you draw your conclusions.

HOMEWORK EXERCISES

1. State the rule for each inverse of the following mappings from Z to Z. State whether each inverse is a mapping.
    a. $x \rightarrow x + 1$
    b. $x \rightarrow 2x$
    c. $x \rightarrow 2x + 1$
    d. $x \rightarrow x - 2$
    e. $x \rightarrow \dfrac{1}{4} x$
    f. $x \rightarrow x$

2. For each of the following mappings give the inverse in the same graphic style. State whether each inverse is a mapping.

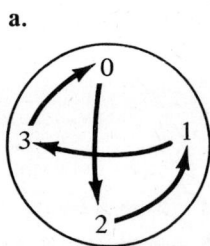

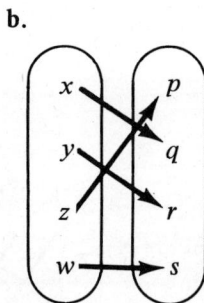

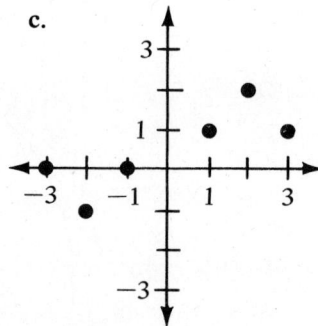

a.   b.   c.

FIGURE 5.32

d. $\{(2, 4), (3, 9), (4, 16), (5, 25), (6, 36)\}$

5.5 COMPOSITION OF MAPPINGS

In Figure 5.33 the rectangle has three distinct symmetries—flips about vertical (V) and horizontal (H) axes, and half-turn (R) about the center point. These rigid motion symmetries can be described as mappings by listing the vertex/image pairs for each motion.

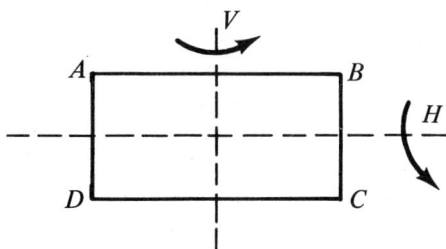

V	H	R
$A \longrightarrow B$	$A \longrightarrow D$	$A \longrightarrow C$
$B \longrightarrow A$	$B \longrightarrow C$	$B \longrightarrow D$
$C \longrightarrow D$	$C \longrightarrow B$	$C \longrightarrow A$
$D \longrightarrow C$	$D \longrightarrow A$	$D \longrightarrow B$

FIGURE 5.33

PROBLEM 1   List the vertex/image assignments that would result in the following:
    a. Motion V is followed by motion H;
    b. Motion V is followed by motion R;
    c. Motion R is followed by motion H;
    d. Motion R is followed by motion R.

The mappings of Problem 1, resulting from performing two assignments in succession, are called *composite* mappings (suggested by one mapping composed of two parts). The composition is symbolized as follows:

$$V \text{ followed by } H \qquad V \circ H$$
$$R \text{ followed by } H \qquad R \circ H$$
$$R \text{ followed by } R \qquad R \circ R$$

PROBLEM 2   Suppose three mappings of Z to Z have rules:

$$x \overset{F}{\to} 2x \qquad \text{(doubles each input)}$$

$$x \overset{G}{\to} x + 1 \qquad \text{(adds one to each input)}$$

$$x \overset{H}{\to} x^2 \qquad \text{(squares each input)}$$

Find the algebraic rule for each composite called for in the following examples. In each case it might be good strategy to check what the composite does to several specific numbers before working out a general rule.

a. $x \overset{F \circ G}{\to} ?$ **Hint:** $3 \overset{F}{\to} 6$ and $6 \overset{G}{\to} 7$ $^-4 \overset{F}{\to} {}^-8$ and $^-8 \overset{G}{\to} {}^-7$

b. $x \overset{F \circ H}{\to} ?$

c. $x \overset{G \circ F}{\to} ?$

d. $x \overset{H \circ G}{\to} ?$

Composition of mappings is a binary operation on mappings, combining two to obtain one. As an operation, the basic properties of closure, associativity, commutativity, identity, and inverse can be checked.

PROBLEM 3   a. First examine closure. Take the two mappings V and H in Problem 1. Is composition of V and H a mapping?

b. Take any two mappings in Problem 1. Is the composition of any two mappings in Problem 1 a mapping?

The composition of mappings in Problem 1 is closed. But the three mappings defined were not sufficient. An additional mapping, I, must be named in order for the composition of R with R to have a result. This is the familiar identity mapping.

$$I$$

$$A \rightarrow A$$
$$B \rightarrow B$$
$$C \rightarrow C$$
$$D \rightarrow D$$

In general, the identity mapping is the mapping of each element of the set to itself. The identity property can be stated as follows: There is an identity mapping I such that $F \text{ o } I = F = I \text{ o } F$ for all mappings F.

PROBLEM 4   **a.** What is the identity mapping for the algebraic rule mappings defined on Z in Problem 2?
   b. Is the composition of algebraic rule mappings defined on Z closed in Problem 2? Explain your answer.

PROBLEM 5   **a.** Is the composition of mappings in Problem 1 commutative? Test several examples.
   b. Is the composition of mappings in Problem 1 associative? Test several examples.

PROBLEM 6   **a.** Is the composition of mappings in Problem 2 commutative? Test several examples.
   b. Is the composition of mappings in Problem 2 associative? Test several examples.

PROBLEM 7   **a.** Does every mapping in Problem 1 have an inverse that is a mapping?
   b. Does every mapping in Problem 2 have an inverse that is a mapping?

Problem 7 confirmed that not all mappings will have inverses that are mappings. The negative one exponent is used as notation for the inverse of a mapping in the same manner as it is used for multiplication inverses. The inverse property can be stated as follows: For each mapping, F, there is a mapping F^{-1} such that $F \text{ o } F^{-1} = I = F^{-1} \text{ o } F$.

★PROBLEM 8   Now consider the composition of mappings defined on any set.
   a. Is the composition of mappings closed?
   b. Is the composition of mappings commutative?

c. Is the composition of mappings associative?
d. Is there an identity mapping?
e. Does every mapping have an inverse that is a mapping?

HOMEWORK EXERCISES

1. Let mapping F be given by Figure 5.34.

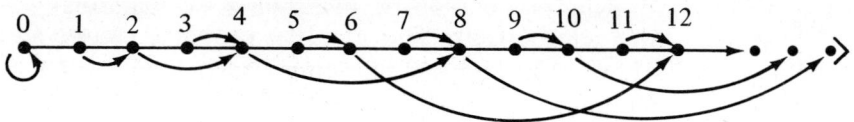

FIGURE 5.34

Let mapping G be given by Figure 5.35.

FIGURE 5.35

Draw a diagram for the composite mapping $F \circ G$.

2. Let P, Q, and R be mappings from Z to Z with the following rules:

$$\overset{P}{x \to 2x + 1} \qquad \overset{Q}{x \to 1 - x} \qquad \overset{R}{x \to 3 \cdot x}$$

a. Find the image of ⁻7 under each of these mappings:
    $P \quad Q \quad R \quad P \circ Q \quad P \circ P \quad Q \circ R \quad P \circ R \quad R \circ P \quad R \circ Q \quad R \circ R$
b. Find the image of x under each of these mappings from part a.
    $R \circ P \qquad\qquad R \circ Q \qquad\qquad R \circ R$
c. What is the image of x under $(P \circ Q) \circ R$? under $P \circ (Q \circ R)$?
d. Is part c an illustration of associativity for mappings?
e. For each of the following, find the required element of Z, if possible.
    If not possible, tell why not.
    1. The integer whose image under P is 13.
    2. The integer whose image under Q is 13.
    3. The integer whose image under $P \circ Q$ is 20.
    4. The integer whose image under $Q \circ R$ is ⁻147.

5. The integer whose image under $R \circ P$ is 19.

6. The integer whose image under $R \circ Q$ is $^-10$.

7. The integer whose image under $P \circ R$ is $^-7$.

f. Write algebraic rules for these mappings:

$$P^{-1} \qquad\qquad Q^{-1} \qquad\qquad R^{-1} \qquad\qquad P \circ P^{-1}$$

3. Repeat Exercise 2 if $x \overset{P}{\to} 2x$, $x \overset{Q}{\to} 1 + x$, $x \overset{R}{\to} 3x - 1$.

4. The following message was coded using the formula $3x + 2$.

$$\text{XQXXQDURC} \qquad \text{XCBBE}$$

The code was constructed in this way. First the alphabet letters were associated with the numbers 1 to 26 in order. Then each letter was converted to a different letter by applying the formula to the associated numbers. For example:

$$A \to 1 \to (3 \cdot 1 + 2) = 5 \to E$$

$$B \to 2 \to 8 \to H$$

a. What is the formula for decoding?

b. What is the message?

5. Code your name using the formula $2x + 1$.

SUMMARY

The concept of mapping or function is a pervasive one in mathematics. In fact, the remainder of this text is devoted to this concept. Mappings are the basis of the definition of congruence, similarity, measure, probability, and statistic. Only the relation is a more fundamental notion. Mapping, a special kind of relation, enters your experience every time you make assignments of number, names, or geometric figures. In efforts to organize some of the data from everyday life, mappings are illustrated through graphs and tables. Even a binary operation is a mapping of an ordered pair to a single element. All of the concepts of mathematics included in this text can be thought of in terms of sets, relations, mappings, and binary operations. When you composed two mappings, you were able once again to examine the abstract structure of another set—this time a set of mappings.

Congruence of Shapes

6

6.1 SAME SIZE AND SHAPE

Pictured in Figure 6.1 are the outlines and pieces of a Middle East map puzzle. Before reading further, use your knowledge of geography or innate spatial intuition to match each country with its location on the map.

While some of the countries, like Israel, Saudi Arabia, or Lebanon, might have been easy to locate from memory of many news reports, you probably had to use common puzzle fitting strategies for others. You mentally moved a piece from its given position to various possible locations on the map outline; or you might even have cut the pieces out from the page and moved them physically to different positions. What you were searching for in either case was a perfect fit between the *size and shape* of puzzle piece and space in the map outline; you wanted to match *congruent* figures by mental or physical *rigid motion*.

This process of matching shapes of objects is extremely common and important in practical situations: nuts are matched to holes in bolts, insides of shoes are matched to feet, doors are sized to people's dimensions, etc.

PROBLEM 1 List a dozen other familiar situations in which the sizes and shapes of pairs of objects must be matched. Then list several situations in which

210

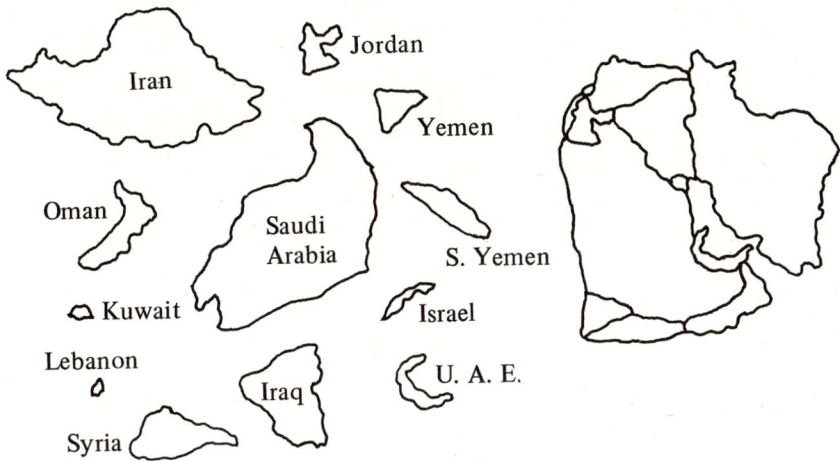

FIGURE 6.1

congruence of shapes is decidedly undesirable. For instance, in panning for gold you don't want nuggets the same shape and size as the sieve holes!

PROBLEM 2   Examine the geometric shapes in Figure 6.2, and sort the shapes into sets whose elements are congruent to each other.

You have looked at some specific examples of rigid motions in previous work. In Section 3.5, for example, you looked at turns and flips restricted to certain geometric figures and generated examples of groups. In Sections 5.1 and 5.2, you examined point and line symmetries that, again, involved the notion of rigid motion with attention focused on specific figures.

PROBLEM 3   Use your intuitive ideas of rigid motion derived from earlier sections to do the following problems.
  a. For each pair of figures you indicated were congruent in Problem 2, name a rigid motion that you think would map one figure to the other.
  b. Choose a pair of figures you think satisfies the relationship: "One figure is the image of the other for a line reflection," and sketch in the line of reflection.

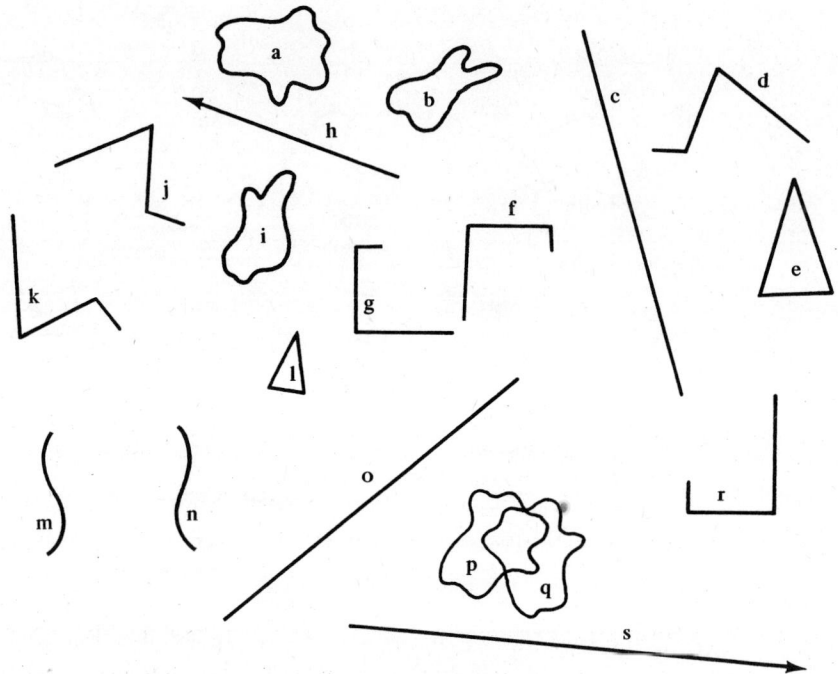

FIGURE 6.2

   c. Choose a pair of figures you think satisfies the relationship: "One figure is the image of the other for a rotation," and locate the point of rotation.

6.2 DEFINITIONS AND UNDEFINABLES IN MATHEMATICS

The problems of Section 6.1 suggest the meaning and importance of congruence in geometry. But how do people know they are thinking of the same properties when they agree that two shapes are congruent? There are several possible ways to communicate this understanding. The simplest is to compare samples of congruent and noncongruent shapes—indicating by example how the terms are to be used. With the vast universe of potential comparisons, this communication by example seems dangerously incomplete.

A second procedure would involve writing in precise detail the defining criteria to be used in checking for congruence. This procedure, commonly associated with rigorous, logically organized mathematics, is not free of difficulties as the discussion in Figure 6.3 indicates:

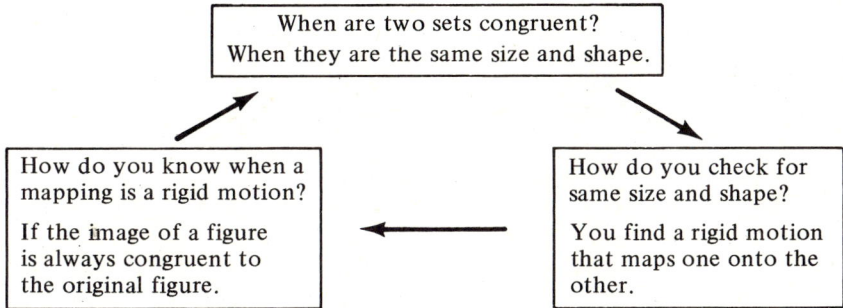

When are two sets congruent?
When they are the same size and shape.

How do you know when a mapping is a rigid motion?

If the image of a figure is always congruent to the original figure.

How do you check for same size and shape?

You find a rigid motion that maps one onto the other.

FIGURE 6.3

PROBLEM 1   a.   What is the role of a definition in ordinary language? What assumptions, if any, underlie the use of a dictionary?

b.   What is the role of definition in mathematics? Consider definitions you have already encountered like *commutative property, group, triangle, factor,* or *relation.* Also look back at how definitions were used in proofs. (See Sections 3.10, 3.12, and 4.10.)

c.   What are the unescapable limitations on defining terms in ordinary language? In mathematics? How do the limitations apply to the circular discussion on congruent figures in Figure 6.3?

Talk over your ideas with the instructor before moving ahead into congruence.

HOMEWORK EXERCISES

1.   Suppose someone told you each pair of the figures in Figure 6.4 is the "same size and shape." In what sense would each claim be justified?

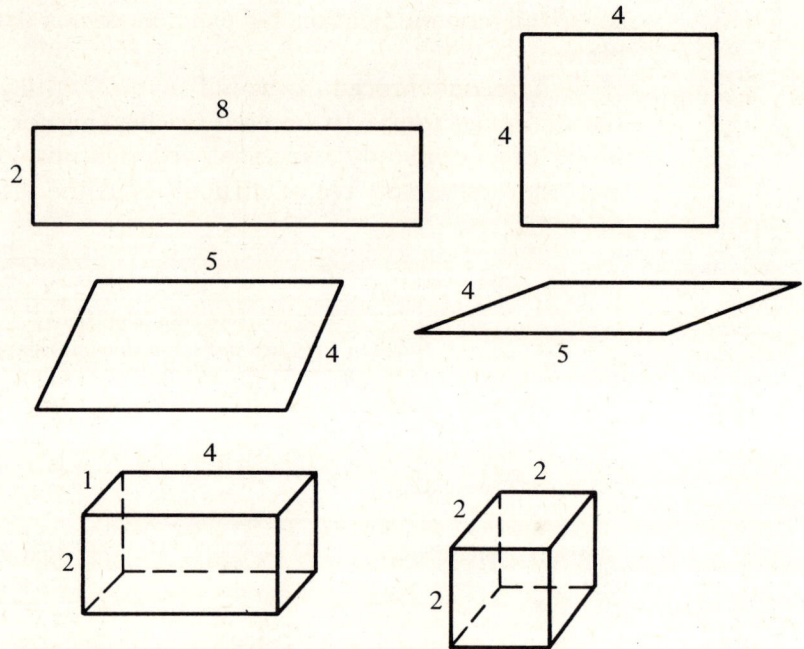

FIGURE 6.4

2. Would you agree that right and left gloves of a pair are the same size and shape? If so, what rigid motion would carry one onto the other? If not, why not?

For a charming story illustrating the mathematical problem of defining *all* of one's terms in discourse, read J. M. Synge's short story "Euclid and the Bright Boy" in *The Mathematical Magpie*, Clifton Fadiman (ed.).

6.3   A GLOSSARY OF MATHEMATICAL TERMS

The problems of the previous section should have indicated the difficulty involved in finding a starting point for a definition. If no words have been previously defined, then what words can you use to define the "next" term? Mathematicians, in their quest for order and lack of circularity, employ the technique of starting with *undefined terms* as a basis for subsequent definitions. Undefined terms are usually those concepts considered so basic and fundamental that there exists some universally accepted notion of the meaning of the term.

Geometrical terms that usually are accepted as undefined are point, line, plane, space, and betweenness; but, there is no rule governing the choice or number of undefined terms useful for a specific purpose.

Table 6.1 is designed to organize the sequential development of a list of geometric terms and symbolic notation, to be used in subsequent sections. Any such list is somewhat arbitrary and you will probably modify the list as well as add to it as you progress. Some of the entries not already included will be determined through discussion problems or in class problems.

PROBLEM 1   To show the role of choice in deciding which terms to leave undefined:
    a. Write a definition of betweenness assuming *point, line,* and *line segment* are undefined or previously defined.
    b. Write a definition of line assuming you have *point, betweenness, line segment,* and *ray* to use.

PROBLEM 2   Attempt to write definitions of *angle, parallel lines, perpendicular lines, distance, polygon,* and *triangle,* in order, using only previously defined terms or undefined terms. Before entering your definitions in the chart, be sure each is compatible with the rest of the class.

6.4   A DICTIONARY OF RIGID MOTIONS

The problems of Sections 6.2 and 6.3 should have demonstrated the futility of ever fully defining terms without circularity unless we assume some concepts as undefined. A promising way to clarify the meaning of *same size and shape, congruent,* and *rigid motion* is to examine the types of motion that seem to preserve size and shape.

A mathematical model of a shape is a set of points in a plane or space. (You will concern yourselves only with rigid motions of a plane in what follows.) The model of a rigid motion is thus a mapping whose domain is the set of points in a plane. Although interest in congruence focuses attention to the images of specific

TABLE 6.1

Term	Symbolic Representation	Definition and/or Basic Characteristic
Point	Capital Letter X, A, Q, etc.	Undefined
Line (straight)	$\overleftrightarrow{AB}$ or k, etc.	Undefined (A set of points)
Plane	plane α	Undefined (A set of points)
Betweenness	X between A and B	Undefined (Relationship of points)
Line segment	$\overline{AB}$	$\overline{AB} = \{A, B\} \cup \{$all points between A and $B\}$
Ray	$\overrightarrow{AB}$	$\overrightarrow{AB} = \overline{AB} \cup \{$all points X such that B is between A and $X\}$
Angle	$\angle ABC$	
Parallel lines	$k \parallel m$	
Perpendicular lines	$k \perp m$	
Distance		
Directed segment	$\overrightarrow{AB}$	
Simple Closed Curve		
Polygon		
Triangle	$\triangle ABC$	
Quadrilateral		
Parallelogram		
Rectangle		
Square		
Rhombus		

figures to see if the rigid motion preserves size and shape, it must be emphasized that *a rigid motion of the plane maps every point of the plane to an image point.*

The problem then is to describe rules for mappings corresponding to our intuitive ideas of a rigid motion and congruence. In earlier work with symmetries, you have encountered several likely candidates: *line reflections* (flips), *rotations* (turns), and *translations* (slides).

You will begin with an investigation of procedures to locate image points using some simple tools like a straight edge, compasses, and tracing paper to solidify our intuitive notions of how the mappings work. Then you will attempt to carefully define the rigid motions.

PROBLEM 1   Using only a piece of tracing paper and a pencil, make a drawing of the set of points shown in Figure 6.5, and the set of image points under reflection in line *k*. Label the image of point *A* as *A'*, the image of point *B* as *B'*, etc. List the steps necessary to locate the image points using the tools specified.

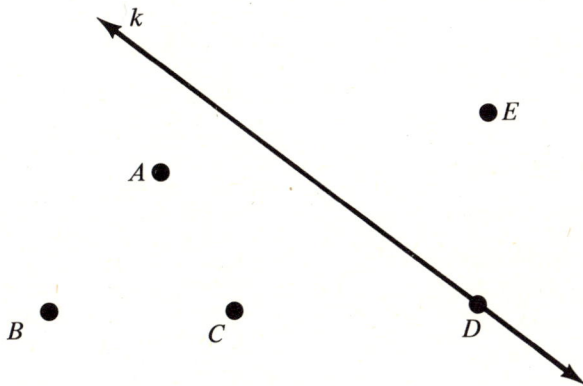

FIGURE 6.5

PROBLEM 2   For each rigid motion mapping described in Figure 6.6, use only the tools specified and follow the directions of Problem 1.
   a. Tools: tracing paper and pencil. Rigid Motion: a rotation about point *P* through the indicated angle.

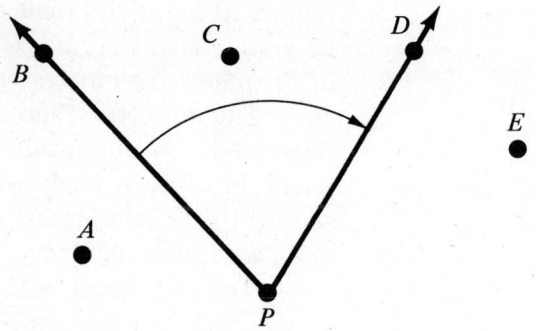

FIGURE 6.6

b. Tools: tracing paper,
   pencil, straight edge.
   Rigid Motion: a half-turn
   at *Q*.

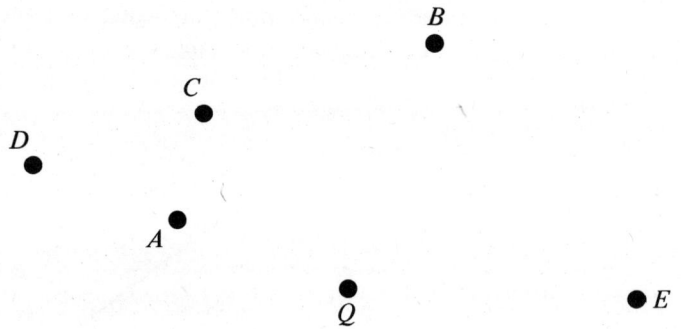

FIGURE 6.7

c. Tools: straight edge,
   compass, pencil. Rigid
   Motion: same as *b*.

FIGURE 6.8

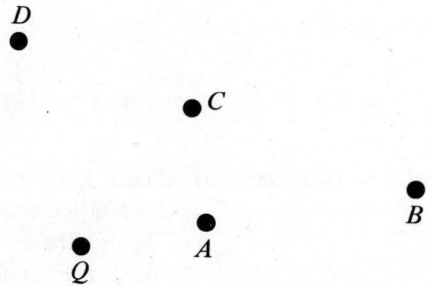

d. Tools: straight edge, tracing paper, pencil. Rigid Motion: a slide in the direction and distance indicated by directed segment $\overrightarrow{AB}$.

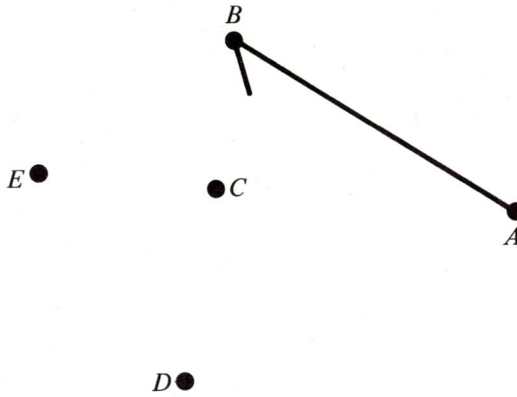

FIGURE 6.9

e. Tools: compass, pencil. Rigid Motion: same as **d.**

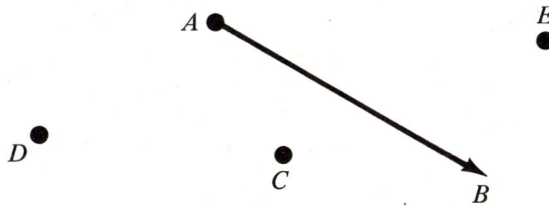

FIGURE 6.10

★f. Tools: compass, pencil. Rigid Motion: line reflection as in Problem 1.

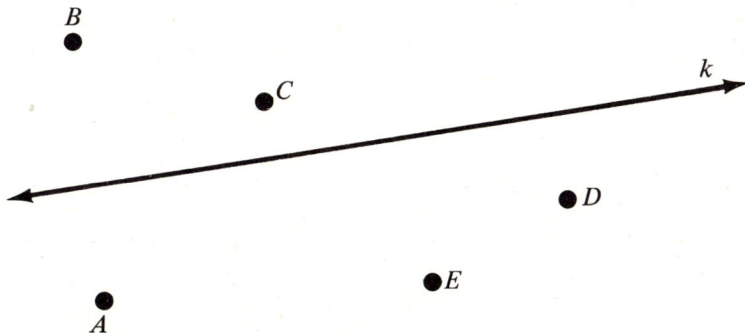

FIGURE 6.11

★g.   Tools: compass, pencil. Rigid Motion: A clockwise rotation about point *C* through an angle determined by ∡ *PQR*.

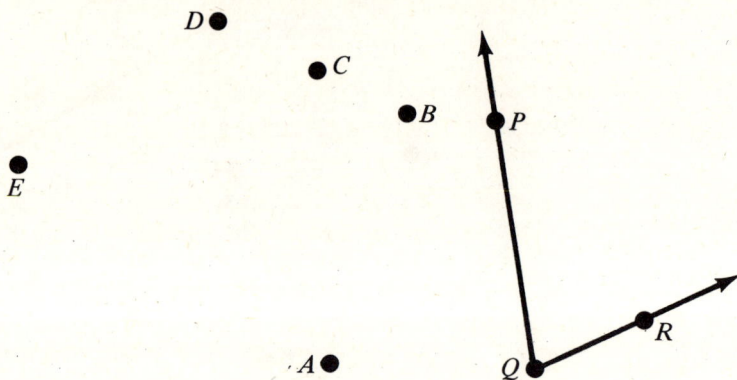

FIGURE 6.12

Hopefully, the preceding problems have sharpened your notions of the effects of line reflections, rotations, and translations. The next set of homework exercises will look at the effects of these motions when applied to a coordinate plane.

HOMEWORK EXERCISES

On a piece of graph paper draw rectangular axes and locate the points *A* (2,6), *B*(4,5), *C*(2,4), *D*(2,1). Then, connect the points with line segments in the order *DCABC.* Use this design in Exercises 1 to 4.

1. Locate and label with coordinates the images of points *A, B, C, D* under reflection in the *x*-axis. Label the images *A′, B′*, etc.

2. Locate and label *A″, B″*, etc., the images of points *A, B, C, D* under reflection in the *y*-axis.

3. Locate and label *A*, B**, etc., the images of points *A, B, C, D* under half-turn about the point (0,0).

4. Locate and label *A**, B***, etc., the images of points *A, B, C, D* under a slide "three right and two up."

5. Study the pattern of image assignments in Exercises 1 to 4, and determine algebraic rules for rigid motions of the coordinate plane:
    a.   reflection about x-axis: $(x,y) \longrightarrow (\ ,\ )$?
    b.   reflection about y-axis: $(x,y) \longrightarrow (\ ,\ )$?
    c.   half-turn about $(0,0)$: $(x,y) \longrightarrow (\ ,\ )$?
    d.   slide to the right of **a** and up **b**: $(x,y) \longrightarrow (\ ,\ )$?

6. Find an algebraic rule for the rotation that maps the coordinate plane $90°$ clockwise about $(0,0)$: $(x,y) \longrightarrow (\ ,\ )$?

7. Find an algebraic rule for the reflection in the line through $(0,0)$ at $45°$ angle to x- and y-axes: $(x,y) \longrightarrow (\ ,\ )$?

PROBLEM 3   Refer to your results of Problem 2. Focus your attention to the images of $\overrightarrow{BC}$ and $\triangle ABC$.
    a.   Point D was always on $\overline{BC}$. Was D' always on $\overline{B'C'}$?
    b.   Does $\triangle A'B'C'$ seem to be congruent to $\triangle ABC$?
    c.   Find images of as many more points as necessary to determine the following:

If you map the points of a plane with a rigid motion (line reflection, rotation, slide) then the set of image points for:

a line is a(n) _____ .
a segment is a(n) _____ .
a ray is a(n) _____ .
an angle is a(n) _____ .

6.5   DEFINING CHARACTERISTICS AND NOTATION FOR RIGID MOTIONS

Each type of rigid motion is determined by certain geometric figures. For example, we must know the line of reflection in order to find the image of a point for a reflection. Similarly, you must know certain specifics for other rigid motions.

PROBLEM 1   Complete the second column of Table 6.2 indicating what information is needed to identify a specific rigid motion of each type.

TABLE 6.2

Rigid Motion	Defining Characteristics
1. Line Reflection	
2. Rotation (clockwise) a. Non-half-turn	
b. Half-turn	
3. Slide	

Note that a half-turn is a specific rotation, thus they are treated separately. Half-turns will receive some special attention later. You will also restrict all rotations from now on to a clockwise direction for simplicity. Any counter-clockwise rotation is equivalent to some clockwise rotation.

The problem of reference to specific rigid motions in a concise notation is quickly solved once you have established the defining characteristics of each kind of rigid motion. Check your results for Problem 1 before continuing on to Problem 2.

PROBLEM 2   Translate the following symbols into a word description of the rigid motion. For the remainder of the chapter, we will use these symbols.

L_k

$R_{C, \measuredangle PQR}$

H_P

$S_{\overrightarrow{AB}}$

HOMEWORK EXERCISES

Carefully locate the images of points X and Y for each rigid motion listed and label the image points as indicated. Use any of the following tools: straight edge, compass, tracing paper, pencil.

1. $L_l\,(X', Y')$

2. $R_{P, \measuredangle ABC}\,(X'', Y'')$

3. $H_Q(X^*, Y^*)$

222

4. $S_{\overrightarrow{FG}}(X^{**}, Y^{**})$

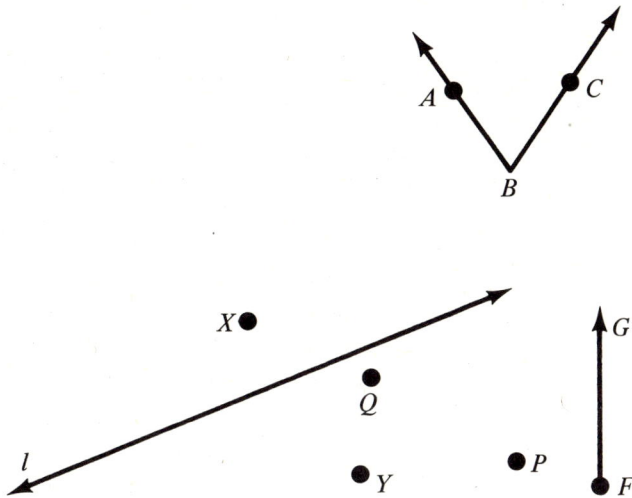

FIGURE 6.13

6.6   DEFINING THE IMAGE POINT

Now you have some idea of how line reflections, rotations, and slides work. You have established the specific defining characteristics of each, and have a notational scheme for representing these rigid motions. You can now turn your attention to carefully defining the relationship between a point and its image for each rigid motion.

PROBLEM 1   **a.** Look back at the examples of line reflections. Review the procedure for locating image points you established in Section 6.4. Based on your observations, complete the following statement so that the image point is completely described.

A line reflection in line k is a mapping of a plane that assigns to each point X of the plane an image point X' such that _____.

Hint: Look at $\overline{XX'}$ relative to k.

b. Write the statement in mathematical notational form.

$$X \xrightarrow{L_k} X' \text{ such that } \underline{\hspace{4cm}}.$$

PROBLEM 2   **a.**   Follow the same procedures as in Problem 1 to complete the definitions below.

$$X \xrightarrow{H_P} X' \text{ such that } \underline{\hspace{4cm}}.$$

$$X \xrightarrow{R_{C,\, \angle PQR}} X' \text{ such that } \underline{\hspace{4cm}}.$$

$$X \xrightarrow{S_{\overrightarrow{PQ}}} X' \text{ such that } \underline{\hspace{4cm}}.$$

b. Construct a summary chart of rigid motions as Table 6.2 in Problem 1, Section 6.5. You will need four columns with lots of space. Label the first column *Notation* and list a specific example such as L_k in row 1.

This completes the task of *describing* the motions. Now look at the special properties of each type of rigid motion and then consider what happens when you compose two rigid motions.

6.7   SPECIAL POINTS AND LINES

Perhaps you noticed during the development of careful descriptions for line reflections, rotations, and slides that some points map to themselves (i.e., $X = X'$), some lines map to themselves ($k = k'$), and some lines map to lines that are parallel ($k \parallel k'$).

PROBLEM 1   **a.**   Add a column to your summary chart for rigid motions. Label the column: *Points That Map to Themselves.* For each type of rigid motion, consider the specific one named as the example in the Notation column and describe those points which map to themselves for that specific motion.

b. Add columns for *Lines That Map to Themselves* and *Lines That Map to Parallel Lines.* Investigate each type of motion as you did in part **a**, and complete the two new columns.

c. If $k = k'$ and $X \in k$, is it ever true that $X = X'$? When?

d. Can a line map to itself in a rigid motion without any points on that line mapping to themselves? Explain.

e. Suppose you define two lines to be parallel if and only if they have no points in common. Does that change any of the entries in your summary chart? How?

HOMEWORK EXERCISES

For each of the following figures, find the image of the figure after the composition of the two rigid motions indicated. Composition of rigid motions is performed just like composition of other mappings: Find the image of the figure for the first motion and then find the image of that *image* for the second motion.

1. $H_P \circ H_Q$

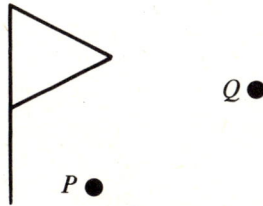

FIGURE 6.14

2. $S_{\overrightarrow{AB}} \circ S_{\overrightarrow{CD}}$

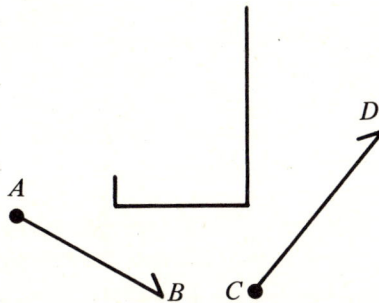

FIGURE 6.15

3. $L_k \circ L_m$

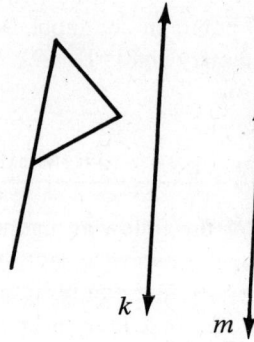

FIGURE 6.16

4. $L_r \circ L_s$

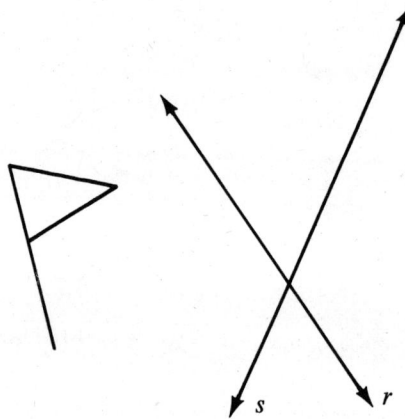

FIGURE 6.17

5. $L_{\overleftrightarrow{AB}} \circ S_{\overrightarrow{AB}}$

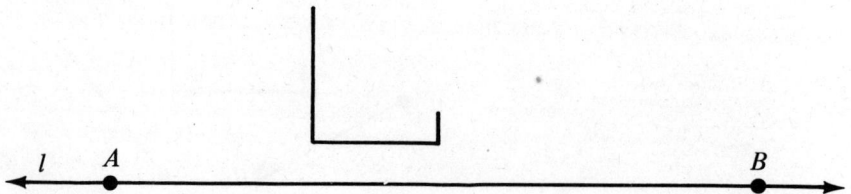

FIGURE 6.18

6.8   A SYSTEM OF RIGID MOTIONS

The preceding set of homework exercises was designed to investigate the idea of composing rigid motions. You should have found that composing two rigid motions yields a single rigid motion, at least for Exercises 1 to 4. Exercise 5 probably still has you stumped trying to find a single rigid motion equivalent to the composition. The specific question, then, is this: Is there one rigid motion equivalent to the composition of a line reflection and slide? A more general question would be: If two figures are congruent, is there a single rigid motion that will map one figure to the other?

PROBLEM 1   Re-examine your results in Homework Exercise 5, Section 6.7. Look carefully at the given figure and its image under the composition $L_{\overrightarrow{AB}} \circ S_{\overrightarrow{AB}}$.

  a. Why is it impossible for $L_{\overrightarrow{AB}} \circ S_{\overrightarrow{AB}}$ to be a line reflection?
     Hint: What do you know about the relationship of the line of reflection and *each* segment $\overline{XX'}$?
  b. Why is it impossible for $L_{\overrightarrow{AB}} \circ S_{\overrightarrow{AB}}$ to be a rotation?
     Hint: Consider points that map to themselves.
  c. Consider the set of all line reflections, rotations, and slides of a plane. What do the results tell us about this set?

The results of Problem 1 indicate that the set of all line reflections, rotations, and slides of a plane is not closed. With only reflections, rotations, and slides to work with, the composition of two of these rigid motions is not always a single rigid motion in the set. Thus, our set of rigid motions, so far, is not all inclusive.

In an attempt to achieve closure for the set of rigid motions, a fourth type of rigid motion will be added. It will consist of a line reflection followed by a slide along the line of reflection. This type of rigid motion is called a *glide reflection* and will be represented by G.

HOMEWORK EXERCISE

1. Locate the image of each figure for the indicated glide reflection.

a. $G_{\overrightarrow{AB}}$

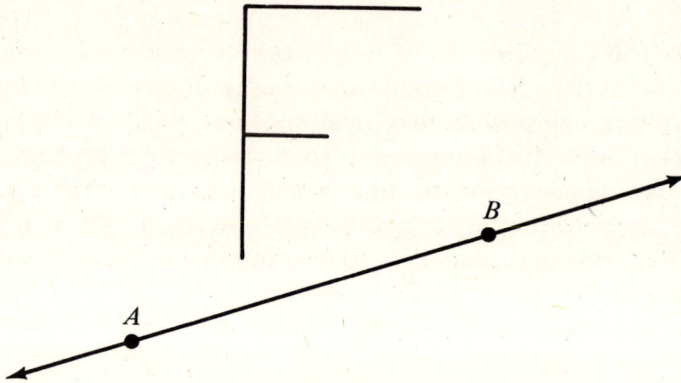

FIGURE 6.19

b. $G_{\overrightarrow{DC}}$

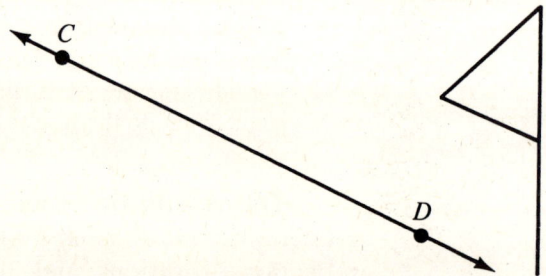

FIGURE 6.20

Let ℜ be the new set of rigid motions that includes glide reflections. Is this set closed under the operation of composition? Problem 2 will look at that question.

PROBLEM 2   Choose a figure, such as flag, ▷, and draw two congruent copies in any position on a sheet of paper. Use tracing paper to try to find a single rigid motion: reflection, rotation, slide, or glide reflection that will map one figure to the other. Repeat the procedure placing the two figures in different positions on the sheet. Is (ℜ, o) closed?

PROBLEM 3   Investigate the structure of ($\mathcal{R}$, o). Is it a group? A commutative group?

HOMEWORK EXERCISES

2. Find inverses for each of the following rigid motions.
   a. H_P
   b. L_k
   c. $S_{\overrightarrow{AB}}$
   d. $R_{P,\, \measuredangle ABC}$
   e. $G_{\overrightarrow{PQ}}$

3. Let $H = \{$ all half-turns of a plane $\}$. Is (H, o) a group? If not, which properties fail?

4. Let $L = \{$ all line reflections of a plane $\}$. Is (L, o) a group? If not, which properties fail?

5. Consider the set of all rotations of a plane. Is it a group under the operation of composition? If not, which properties fail? Will any subset of this set form a group under composition? Explain.

PROBLEM 4   You have discussed the relationship of congruence and rigid motions. You should have found that ($\mathcal{R}$, o) is a group. Congruence is a relation defined on the set of geometric figures. Using the properties of ($\mathcal{R}$, o), show that congruence, as we will define it, is an equivalence relation on the set of geometric figures.

> *DEFINITION:* Figure A is congruent to Figure B if and only if there is a rigid motion which maps Figure A to Figure B.

PROBLEM 5   Using the definition of congruence, write the two implications involved.

★6.9   ANOTHER LOOK AT LINE REFLECTIONS

In Homework Exercises 3 and 4 of Section 6.7 you looked at the composition of two line reflections. You found that the composition of two parallel line reflections was equivalent to a slide and the composition of two line reflections through intersecting lines was equivalent to a rotation.

PROBLEM 1   a. Are all slides equivalent to a composition of line reflections?
b. Are all rotations equivalent to a composition of line reflections? If so, how would you determine the necessary lines of reflection?
c. What does your result to part **a** tell you about glide reflections?
d. Generalize your findings into a general statement relating line reflections to the other types of rigid motions.

HOMEWORK EXERCISES

For each pair of congruent figures in Figure 6.21 locate and label lines of reflection so that one figure can be mapped to the other by a composition of line reflections through the lines you have located. Specify the composition that accomplishes the task.

1.

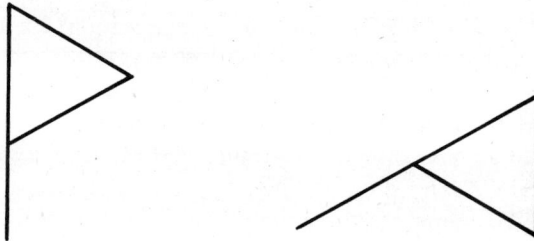

FIGURE 6.21

2.

FIGURE 6.22

6.10   REFLECTING, ROTATING, AND SLIDING TO CONGRUENCE

When two lines intersect, as pictured in Figure 6.23, the opposite angles formed appear to be congruent in pairs. This suspicion can be checked physically by tracing one angle of each pair and trying to reflect, rotate, or slide the tracing onto its proposed congruent mate. The mathematical check for congruence calls for description of a rigid motion mapping that would map ⊰*APB* onto ⊰*CPD* and ⊰*BPC* onto ⊰*DPA*. If you recall the definition and assignment properties of half-turns, it seems likely that this simple mapping will match up the desired angles.

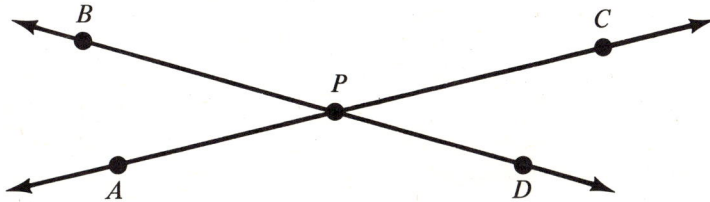

FIGURE 6.23

In the following problems, situations are described. You are to: **(a)** conjecture, on the basis of given information, some congruence relationships, and **(b)** describe the rigid motion or motions that you believe would carry one figure onto its conjectured congruent image.

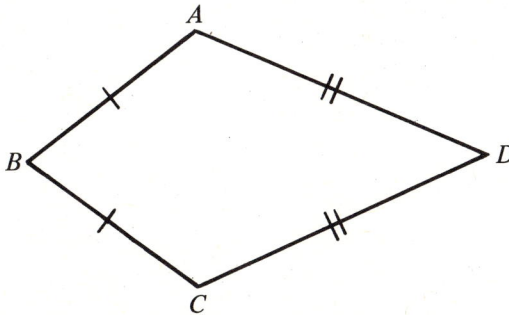

FIGURE 6.24

231

PROBLEM 1   In Figure 6.24, assume that $\overline{AB} \cong \overline{CB}$ and $\overline{AD} \cong \overline{CD}$. Are any other parts of the figure congruent? If so, what rigid motion would confirm it?

PROBLEM 2   In Figure 6.25 $\overline{AB} \cong \overline{CD}$. Are the circles congruent? If so, what rigid motion would confirm it?

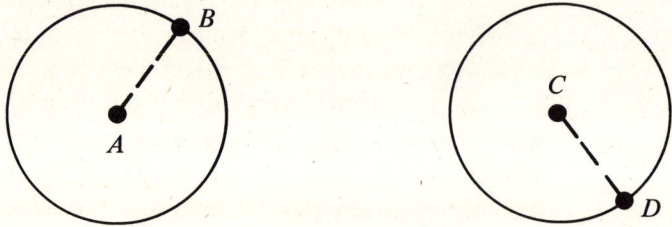

FIGURE 6.25

PROBLEM 3   In Figure 6.26, assume that line m is parallel to line n and line k cuts both. List all pairs of angles you conjecture will be congruent stating the simplest possible rigid motion that you guess would confirm each congruence.

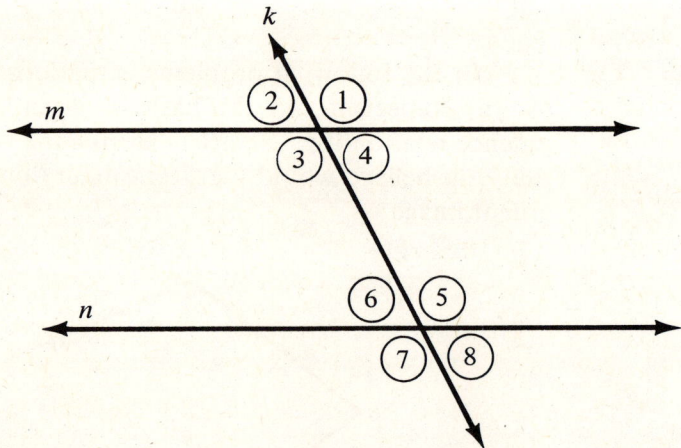

FIGURE 6.26

PROBLEM 4   Assume only that in Figure 6.27, $\overleftrightarrow{AB}$ is parallel to $\overleftrightarrow{CD}$ and $\overrightarrow{AE}$ is parallel to $\overrightarrow{CF}$. Are any congruent angles suggested?

232

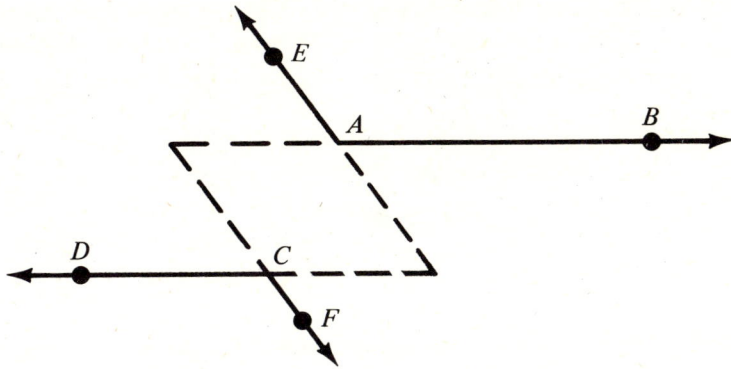

FIGURE 6.27

PROBLEM 5   Assume that in Figure 6.28, $\overline{AB} \parallel \overline{CD}$, and $\overline{DA} \parallel \overline{CB}$. Which sides are congruent? Which angles are congruent? What relationships are formed by the intersecting diagonals?

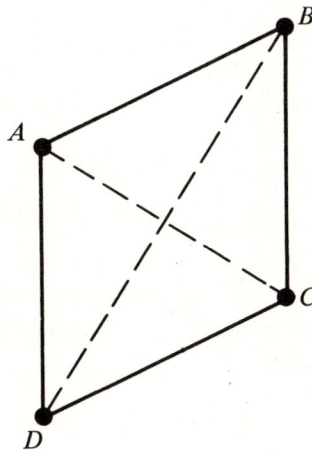

FIGURE 6.28

PROBLEM 6   Assume that in Figure 6.28 the diagonals $\overline{AC}$ and $\overline{BD}$ are perpendicular and bisect each other. What parts of the quadrilateral are congruent?

PROBLEM 7   Assume the triangle in Figure 6.29 has two congruent angles, $\angle A \cong \angle B$. What other parts are congruent? Why?

233

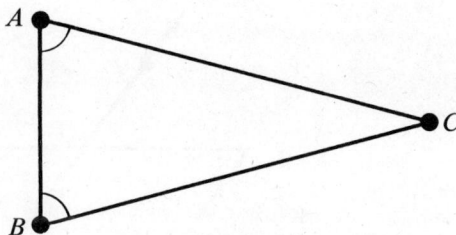

FIGURE 6.29

★PROBLEM 8   Draw several triangles of varying shapes and sizes. Then for each, tear off the corners and place them vertex to vertex and edge along edge as shown in Figure 6.30. What do you notice happening in every case?

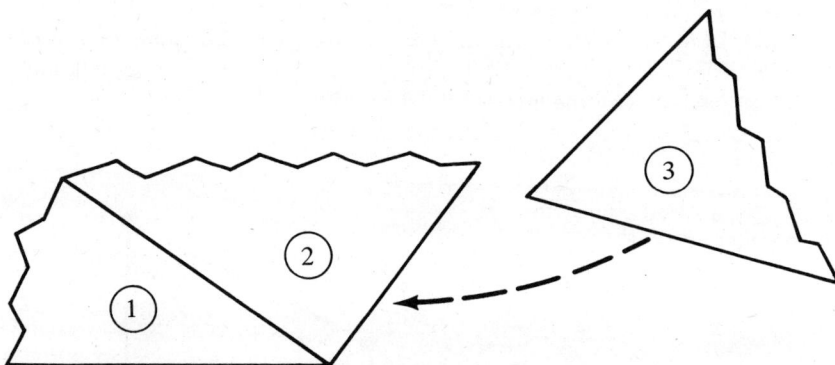

FIGURE 6.30

Can you describe rigid motions that will map the angles of a triangle into such position without the physical tearing? Look at Figure 6.31 for a hint.

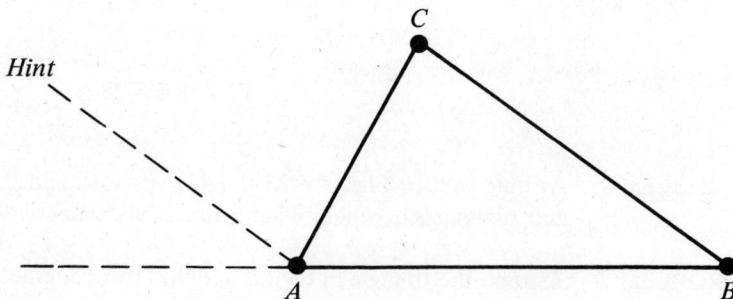

FIGURE 6.31

234

In this section you have begun searching informally for places where congruence of geometric figures occurs regularly. Then you have conjectured the kind of rigid motion mapping that would likely confirm your hunch about congruence. Some of the concepts of congruence and rigid motion will be organized in the next section.

6.11 SMALL SCALE AXIOMATICS—HALF-TURNS

In exploring congruence, you have worked through three phases in the evolution of a mathematical theory. First, you perceived interesting patterns suggesting underlying conceptual similarity in many situations. Second, you attempted to define the concept of interest in order to communicate your idea to others. Third, you observed situations in which certain congruence facts imply congruence relationships.

The fourth phase of development involves organization of facts about congruence into a logical system of specified elements, operations, or relations, properties taken without proof called axioms, and properties proven from axioms agreed upon, called theorems. You have gone through this procedure at least once before, in the analysis of field structure in $(Z_7, +, \cdot)$. In that algebraic system you developed two lists of properties, the fundamental properties and the other properties whose truth could be inferred from the fundamental properties. Although there was some freedom of choice in the basic list (different sets of assumptions could logically lead to the same set of proven facts), mathematicians have generally agreed on a uniform list of eleven field properties.

In this section we will make a modest step toward "axiomatic organization" of the congruence concept and relations observed informally in Section 6.10. The theory of congruence could be organized differently or in more complete form. *The purpose here is to demonstrate the method of mathematics, not to exhaust the domain of facts about congruence.*

UNDEFINED TERM: Same size and shape

DEFINITIONS: Set A is *congruent* to set B, denoted by

$A \cong B$, if and only if A and B are the same size and shape.

A mapping is a *rigid motion* if and only if it maps each set to a congruent image.

AXIOM 1:   A rigid motion maps lines to lines, segments to segments, rays to rays, and angles to angles, i.e., figures to figures of the same type.

Having set down the basic ideas involved in congruence— same size and shape, rigid motion—proceed to organize the facts explained by one type of rigid motion, a half-turn.

DEFINITION: A mapping is a *half-turn* about P if and only if it assigns P to P and each point X, where $X \neq P$, an image X' such that P is on line $\overleftrightarrow{XX'}$ with $\overline{XP} \cong \overline{X'P}$ (i.e., P is the *midpoint* of $\overline{XX'}$).

AXIOM 2:   A half-turn is a rigid motion.

Note that the definition of a half-turn draws only on terms previously introduced to the discourse. The axiom says to followers of the theory of congruence, "We hope you will agree that this is a reasonable assumption and will base later conclusions on it."

THEOREM 1:   If the diagonals of a quadrilateral bisect each other, opposite sides of the quadrilateral are congruent.

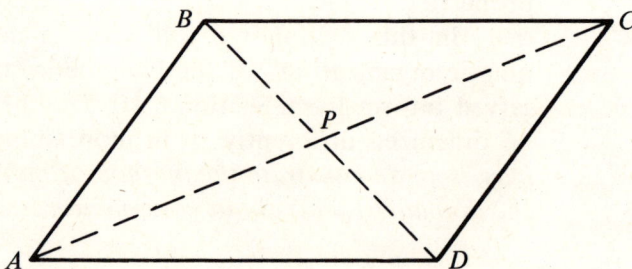

FIGURE 6.32

ANALYSIS OF PROOF: The task is first to describe a rigid motion that maps $\overline{AB}$ onto $\overline{CD}$ and $\overline{AD}$ onto $\overline{CB}$; then to prove that the desired congruences follow. See Figure 6.32.

PROOF: Consider H_P

H_P is a rigid motion by Axiom 2.

Since P is given as the midpoint of $\overline{AC}$ and $\overline{BD}$, the definition of a half-turn guarantees that H_P maps:

$$A \to C$$
$$B \to D$$
$$C \to A$$
$$D \to B$$
$$P \to P$$

Thus, $\overline{AB} \to \overline{CD}$ and $\overline{AD} \to \overline{CB}$ by Axiom 1.

So, $\overline{AB} \cong \overline{CD}$ and $\overline{AD} \cong \overline{CB}$ by the definition of rigid motion.

This completes the argument. Note how it is based on the definition of congruence, rigid motion, and half-turns, and the axioms of rigid motions and half-turns.

PROBLEM 1   ***THEOREM 2:*** If two lines intersect, the angles formed are congruent in opposite pairs.

ANALYSIS OF PROOF: Describe the needed rigid motion and explain precisely why it confirms the desired congruences. See Figure 6.33.

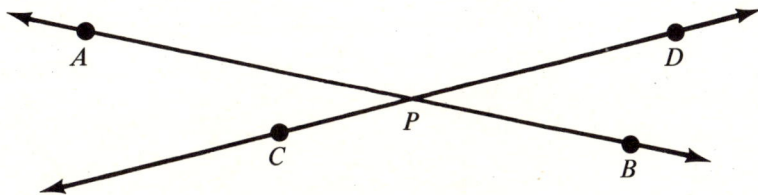

FIGURE 6.33

237

The relationships in Theorems 1 and 2 were actually mentioned in section 6.10 where you were also asked to conjecture a rigid motion carrying figures onto congruent images. What is the difference here? You have been asked to prove that the mappings chosen really do what is hoped for. This aspect of proof structure becomes clearer in the next group of theorems, depending on a special property of half-turns.

AXIOM 3: For any half-turn, the *image* of each line $\overleftrightarrow{XY}$ is a parallel line $\overleftrightarrow{X'Y'}$.

You have already noted this property in Section 6.7. Here the property is stated as an assumed fact to be used in establishing other relationships.

PROBLEM 2   Devise proofs for each of the following theorems about congruence. Follow the model given for Theorem 1 (Figure 6.32).

THEOREM 3: If parallel lines are cut by a transversal, then the alternate interior angles are congruent. Refer to Figure 6.34. ($\angle 1 \cong \angle 3$ and $\angle 2 \cong \angle 4$)

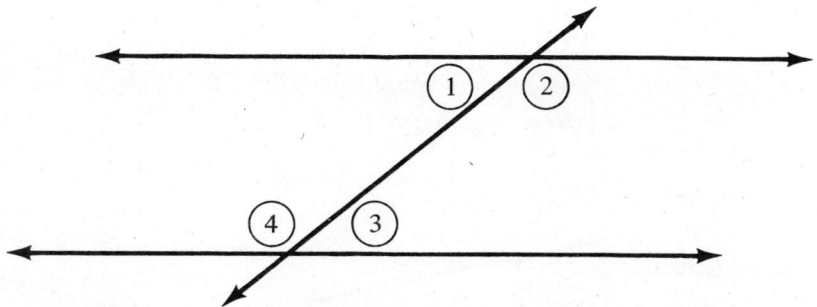

FIGURE 6.34

THEOREM 4: If parallel lines are cut by a transversal, then alternate exterior angles are congruent. Refer to Figure 6.35. ($\angle 1 \cong \angle 3$ and $\angle 2 \cong \angle 4$)

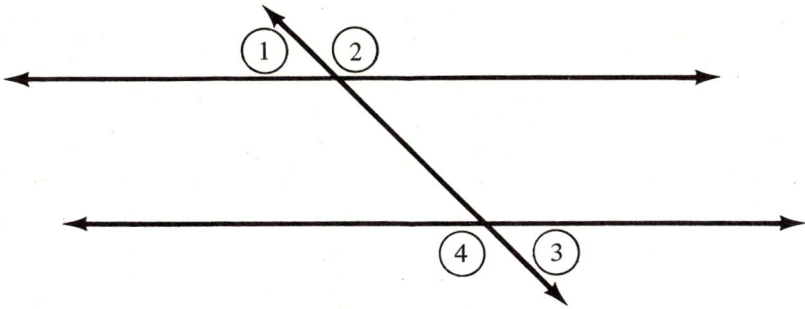

FIGURE 6.35

THEOREM 5: In Figure 6.36, line k is parallel to side $\overline{AB}$ of the triangle. Thus $\angle A \cong \angle 1$ and $\angle B \cong \angle 2$. What familiar fact of geometry does this suggest?

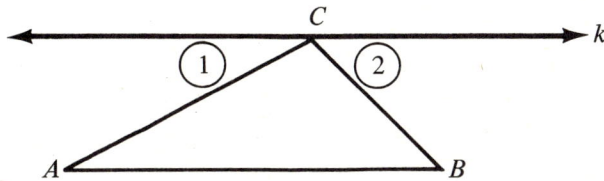

FIGURE 6.36

★**THEOREM 6:** If, as shown in Figure 6.37, a quadrilateral has its opposite sides parallel, the following relationships hold:

a. opposite sides are congruent;

b. opposite angles are congruent;

c. diagonals bisect each other.

Hint: Try a half-turn at the midpoint of $\overline{AC}$. Where must $\overline{AB}$ map? Where must $\overline{CB}$ map?

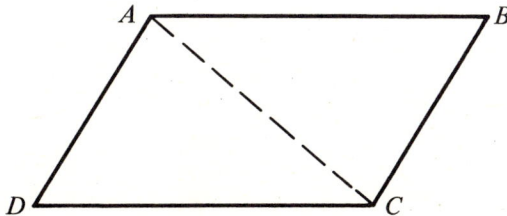

FIGURE 6.37

239

The homework problems will extend this small axiomatic system to new congruence results. You will not be challenged to a similar rigorous development of results depending on reflections or slides, but you will devote the last section to summary analysis of rigid motions and their uses.

HOMEWORK EXERCISES

★1. Suppose in quadrilateral $ABCD$, $\overline{AB}$ is parallel and congruent to $\overline{CD}$. Prove that $\overline{AD}$ is parallel and congruent to $\overline{BC}$.

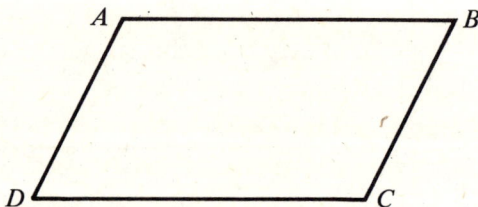

FIGURE 6.38

★2. If you are interested in building a small axiomatic system based on properties of line reflections, try starting with the following.

DEFINITION: Two lines are *perpendicular* if and only if the angles formed at their intersection are all congruent to each other.

AXIOM 1: The *perpendicular bisector* of $\overline{AB}$ is the set of all points X such that $\overline{XA} \cong \overline{XB}$.

DEFINITION: A mapping is a *reflection in line k* if and only if it assigns each point of k as its own image and assigns every other point X an image X' such that k is the perpendicular bisector of $\overline{XX'}$.

AXIOM 2: A reflection in line k is a rigid motion.

Now return to Section 6.10 in search of congruent facts likely to be provable using properties of line reflections.

S U M M A R Y

In this chapter you looked at a special kind of mapping—rigid motion—related to the geometric concept of congruence. As in past chapters, a careful investigation of the properties of the new concept followed the introduction of the concept. You found that the set of rigid motions of a plane (a set of mappings) form a non-commutative group with the operation of composition.

In Sections 6.10 and 6.11 you noted how the properties of rigid motions could be used to prove other more familiar geometric relationships. The next chapter will investigate another common geometric concept: measure. Measure will be looked at as another example of a special kind of mapping, and the properties of the measure mappings will be investigated in relation to familiar problems in geometry.

Measurement

7

7.1 AN OVERVIEW

Measurement is a very integral part of all of our lives. Though you may not stop to consider each measurement situation as it occurs, you continually make decisions in which measurement is involved.

PROBLEM 1 Let's imagine that upon graduation you secure a teaching position in a school some distance away, so you are faced with moving from your present location. You rent an efficiency apartment at a reasonable rate but you must decorate and furnish it yourself.
   a. Select two or three of the items listed below for which you might need to find the cost and fill in Table 7.1. Item 3 has been started as a model.
   1. gasoline needed to drive to your new home
   2. moving your possessions to your new apartment
   3. the cheapest transportation from your apartment to your school
   4. painting apartment walls and ceilings
   5. carpeting the dining-living room area
   6. draperies for the windows
   7. optimum size room air conditioner, if appropriate
   8. optimum size stove and refrigerator
   9. a shower curtain for a tub/shower

TABLE 7.1

Item	Measurement Data That Might Be Needed	Sample Measurement You Might Expect to Obtain
Item Number 3	distance from apt. to work	13.5 kilometers
Cheapest transportation to work	bus fare bus times	.50 8:15 AM, 8:45 AM
Item number ___		
Item number ___		

b. Carefully examine your completed table and search for common properties of, or similarities between, the different kinds of measures and/or measurements you listed. Which measurements are exact? Which measurements are approximations?

c. What general steps or procedures must one go through to *measure*?

d. Why do we measure?

e. Is measurement an example of any of the mathematical structures we have seen previously? A relation? A mapping? An operation?

f. When we measure, do we find the exact measure? Explain.

For more information about the approximate nature of measurement read Harold Trimble's article "Teaching About 'About' " in the February 1973 *Arithmetic Teacher,* National Council of Teachers of Mathematics.

7.2 PROPERTIES OF MEASUREMENT

In Section 7.1 you examined a variety of measures. In each case you assigned a number to some characteristic of the object being measured.

PROBLEM 1   Consider Figure 7.1, a diagram of a regular pentagon (all five sides congruent). One measurement mapping assigns the number called the perimeter to the regular pentagon. What is this perimeter number? The length of a side measured in centimeters is 4.

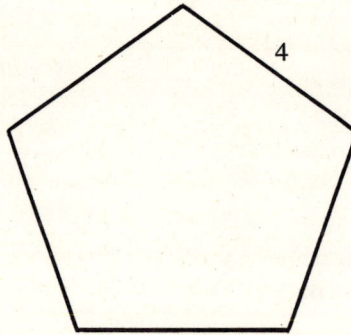

FIGURE 7.1

Thus, measurement can be viewed as a mapping whose range is a set of numbers. Weighing physical objects is a mapping whose domain is the set of objects being weighed and whose range is a set of weights (numbers) labeled with an appropriate unit for reference. When items are priced for distribution in a supermarket, the set of items is mapped to a set of prices (numbers) with an appropriate unit (dollars) either labeled or understood.

PROBLEM 2   For each of the measurement mappings that follows, list the domain and range of the mapping.

	Domain	*Range*
a. length		
b. area		
c. volume		
d. angle measure		

THE COMPARISON PROPERTY: Only measures of the same type can be compared. For example, the distance between two points can be compared to the distance between two other points, but can not be meaningfully compared to the cost of a doughnut at the local pastry shop. We can compare measures of the same type, regardless of the units used, as long as we have some way to compare the relationship between the different units.

The comparison property can be illustrated by the numbers assigned to the lengths of nails. Figure 7.2 pictures some standard size nails.

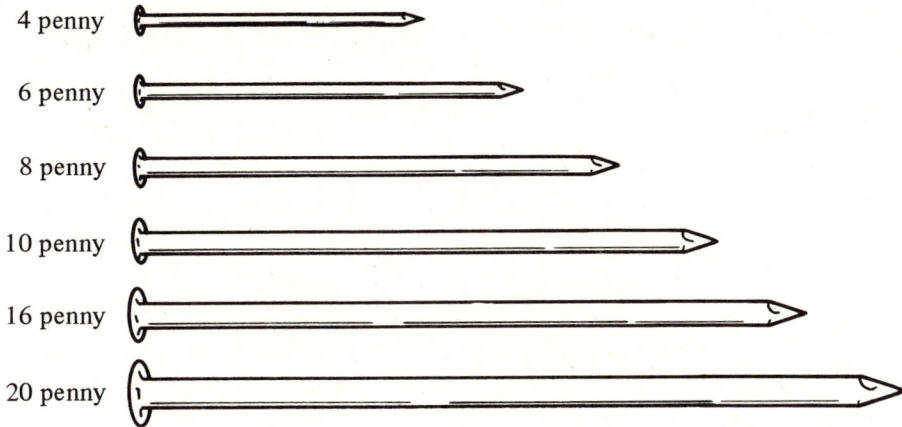

4 penny

6 penny

8 penny

10 penny

16 penny

20 penny

FIGURE 7.2

PROBLEM 3   **a.** Using this measure for nails, determine which of the following statements are true.
   1. An eight penny nail is twice the length of a four penny nail.
   2. A six penny nail is half the length of a twenty penny nail.
   3. The sum of the lengths of a four penny nail and a six penny nail is the same as the length of a ten penny nail.
   4. An x penny nail is longer than a y penny nail if and only if $x > y$.

   b. Write some statements of your own about the relationship of standard size nails.

   c. Discuss the advantages and disadvantages of assigning numbers in this way to nails.

THE COVERING PROPERTY: Any specific example of the property being measured can be used as the unit for measuring that property in other examples. Any line segment can be used as the unit of measure for length. By laying off non-overlapping copies, end-to-end, of any segment designated as a unit, you can eventually cover any other segment with a finite number of copies of the unit. This property applies to area, volume, and angle measure as well.

PROBLEM 4   a. Let be the unit region in measuring the area of the polygon region shown as Figure 7.3. What is the number of unit regions needed to cover the polygon region? The area of this polygon region is _____ unit regions.

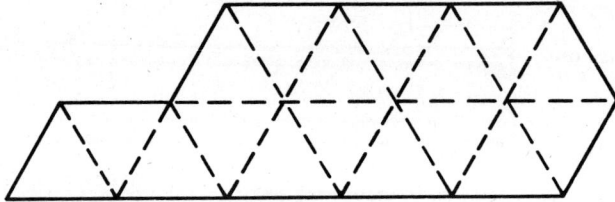

FIGURE 7.3

b. Using as the unit region find the area of the region enclosed by the simple closed curve in Figure 7.4.

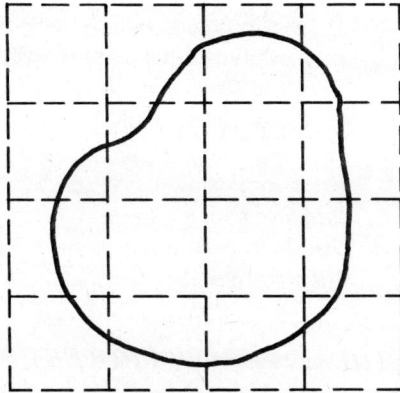

FIGURE 7.4

c. Describe some unit regions that might be suitable for measuring volume.

d. Find the measure of ∡PQR using ∡ABC as the unit angle.

246

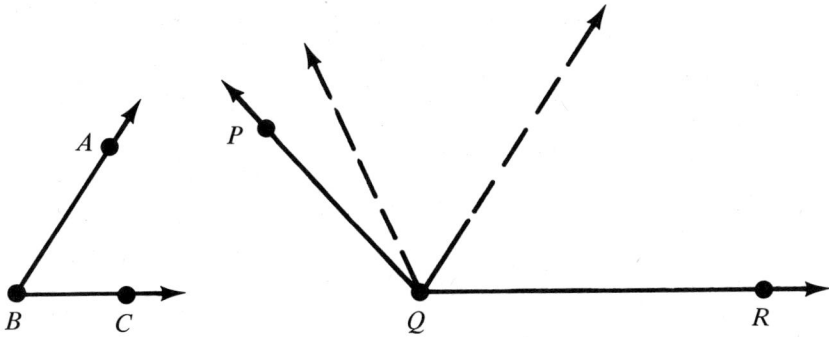

FIGURE 7.5

In Problem 4 you are using arbitrary nonstandard units as covering units. A meter is an arbitrary unit of length measure, but is important because of its universal acceptance as a *standard unit of measure.*

PROBLEM 5   List as many kinds of measurement as you can, along with an appropriate standard unit used for each measure.

THE CONGRUENCE PROPERTY: If 2 sets are congruent, then there exists at least one characteristic such that the measures of that characteristic for the 2 sets are equal. As an example, consider 2 line segments, $\overline{AB}$ and $\overline{CD}$, that are congruent. A characteristic of line segments that would yield equal measures for the two segments is length, $m(\overline{AB}) = m(\overline{CD})$. A natural question one could immediately ask is: If 2 sets are congruent, are all measures of characteristics for the 2 sets equal? Again using your 2 line segments, $\overline{AB}$ and $\overline{CD}$, a characteristic you could measure, other than length, would be the slopes of the segments relative to some coordinate axes. The slope of $\overline{AB}$ can be different from the slope of $\overline{CD}$, even though the segments are congruent. So congruence does not guarantee all measures of the congruent sets will be equal.

PROBLEM 6   a. For each pair of congruent figures described, list some properties

247

of the figures that would yield equal measures for the figures and some properties that would yield different measures. ($\cong$ means "is congruent to.")

	Properties with Equal Measure	Properties with Possible Unequal Measures

1. $\angle XYZ \cong \angle RST$

2. $\triangle ABC \cong \triangle PQR$

3. Prism $ABCDEF \cong$ Prism $PQRSTV$

b. Complete the statements of the following theorems with relationships suggested by your previous explorations. You may find several ways of completing the statements.
  1. If two segments are congruent, then . . . .
  2. If two angles are congruent, then . . . .
  3. If two triangles are congruent, then . . . .
  4. If two prisms are congruent, then . . . .

c. You might now ask if the converse of each relationship established in part **b** is true. If measures are equal, are figures congruent? Write the converse of each statement you completed in part **b** and determine if the converse is true. Justify your conclusion.

THE ADDITIVE PROPERTY: If A, B, and C are sets such that $A \cup B = C$, then the measure of C = measure of A + measure of B providing the measure of $A \cap B$ is zero. As an illustration of this property for linear (length) measure, let $Q \in \overline{PR}$. $\overline{PR} = \overline{PQ} \cup \overline{QR}$ and $m(\overline{PQ} \cap \overline{QR}) = m(Q) = 0$, so $m(\overline{PR}) = m(\overline{PQ}) + m(\overline{QR})$.

How would you find the area measure of polygonal region $ABCDE$ in Figure 7.6? Using the additive property: The area of polygonal region $ABCDE$ will equal the area measure of triangular region ABC + the area measure of rectangular region $ACDE$ since polygonal region $ABCDE$ is the union of triangular region ABC and rectangular region $ACDE$, and the intersection of those two regions, $\overline{AC}$, has an area measure of zero.

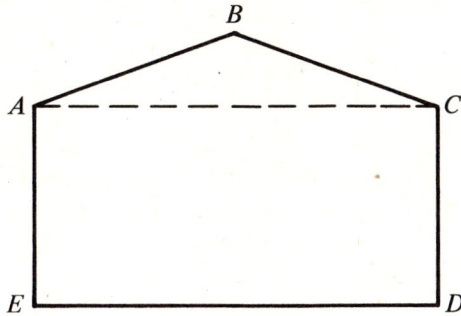

FIGURE 7.6

What the additive property really does is allow you, with certain restrictions, to find a measure of a *whole* by summing the measures of *parts*.

PROBLEM 7   What would happen if the restriction about the intersection of the subsets was not included in the additive property? **Hint:** Look at a segment, $\overline{AB}$, with 2 points, P and Q, on $\overline{AB}$ such that P is between A and Q. Is $\overline{AB} = \overline{AQ} \cup \overline{PB}$? Is $m(\overline{AB}) = m(\overline{AQ}) + m(\overline{PB})$?

HOMEWORK EXERCISES

1. Describe how the additive property of measure is used in each of the following situations:
   a. checking out at a supermarket;
   b. computing a final grade;
   c. buying an air conditioning unit for a home;
   d. buying paint;
   e. planning the size of an auditorium for an elementary school.

★2. Measure can be thought to have a *subtractive property*. Write a statement of what you think the subtractive property would be. Illustrate your statement (definition). Try to prove the subtractive property using the additive property together with the properties of numbers previously developed.

7.3 MEASUREMENT ACTIVITIES

For each of the activities described, do not use any tools specifi-cally designed for measuring, and *keep a record* of any assump-tions and/or decisions made prior to commencing the activity, the steps followed during the activity, and any results or conclusions obtained.

PROBLEM 1
    a. Find the perimeter of your classroom.
    b. Find the surface area of your classroom.
    c. Find the volume of your classroom.
    ★**d.** Find the measures of all the angles formed by rays having an upper corner of the door frame of your classroom as an endpoint and containing a corner point of the wall that contains the door.

PROBLEM 2
To answer the following questions, examine the records kept for the activities you did in Problem 1.
    a. Were there kinds of preliminary assumptions/decisions made that were common to all the activities? If so, what were they?
    b. How were the procedures used in the activities alike?
    c. Were results of one activity useful for other activities? Explain.
    d. Were your results exact? Is there an exact answer to each problem?

PROBLEM 3
For each of the four measure properties—comparison property, congru-ence property, covering property, and additive property—identify where and how the property was used in your progress through the measurement in Problem 1.

HOMEWORK EXERCISES

1. Describe as many different ways as you can to measure a half-dollar. A coffee can. A sheet of paper. A ball. Different unit does not imply a dif-ferent way of measuring.

2. For each kind of measure listed, suggest a common household or school-room item other than standard measuring instruments that could be used as a unit of measure. Briefly describe a procedure you could use to mea-sure with your unit.
    a. length

   b. area
   c. volume
   d. angle measure
   e. weight

3. Consider a 10 by 10 by 10 unit block. What is its volume? Put another one next to it. What is the volume of the rectangular prism formed?

4. Consider a 5 by 5 by 5 unit block. How many blocks make one 10 by 10 by 10 unit block? How many 3 by 3 by 3 unit blocks does it take to make one 6 by 6 by 6 unit block? In general, if you double each side of a cube its volume is increased by a factor of _____ .

7.4 THE INTERNATIONAL SYSTEM OF UNITS (SI)

The International System of Units, referred to as SI, is an internationally accepted set of standard measurement units. It is the modernized metric system established in 1960. The following brief historical review traces the steps toward metrication in the United States.

1790 The metric system of measurement was developed by the French Academy of Science in 1790. The U.S. Congress discussed the need for a uniform system of measurement, but no action was taken toward adopting one.

1866 Legislation in the United States made it "lawful throughout the country to employ the weights and measures of the metric system" but it was not made mandatory.

1875 The Treaty of the Meter was signed by 17 countries including the United States. This treaty set up well defined metric standards for length and mass. The U.S. Bureau of Weights and Measures was established at this time.

1893 The yard was legally defined as a fractional part of a meter and the pound as a fractional part of a kilogram.

1959 All customary measuring units were officially defined in terms of metric units.

251

1968   The U.S. Congress directed the Secretary of Commerce to undertake a three-year U.S. Metric Study.

1971   The Report of the U.S. Metric Study was transmitted to Congress. The report recommended that the United States change to the International Metric System through a coordinated national effort, and that a target date ten years ahead be established.

1972   The Metric Conversion Act was passed by the Senate.

1974   The Elementary and Secondary Educational Act, which passed both houses and was signed into law by the President, contained the following section:

Education For the Use of the Metric System of Measurement.

Under this section the Commissioner of Education is authorized to carry out a program of grants and contracts in order to encourage educational agencies and institutions to prepare students to use the metric system of measurement. $10,000,000 is authorized for this use for each of the fiscal years ending prior to July 1978.

1975   The Metric Conversion Act of 1975 which the President signed on 23 December 1975 became Public Law 94-168. It has the following introduction: "To declare a national policy of coordinating the increasing use of the metric system in the United States, and to establish a United States Metric Board to coordinate the voluntary conversion to the metric system."

Since the United States is already committed to a large scale educational effort to teach the metric system in the elementary schools, it is imperative that both pre-service and in-service early-childhood, elementary, and special education teachers be comfortable and competent using the metric system.

The meter (metre) is the basic SI unit of length. Using familiar Greek and Latin prefixes that unit can be increased or decreased as follows:

1 kilometer = 1000 meters

$$1 \text{ hectometer} = 100 \text{ meters}$$
$$1 \text{ dekameter} = 10 \text{ meters}$$
$$1 \text{ meter} = 1 \text{ meter}$$
$$1 \text{ decimeter} = 0.1 \text{ meter}$$
$$1 \text{ centimeter} = 0.01 \text{ meter}$$
$$1 \text{ millimeter} = 0.001 \text{ meter.}$$

The same prefixes are applied to the basic SI unit of mass (weight), the gram. The gram and the kilogram are the units of interest at the elementary school level.

The basic unit of volume is defined in terms of the cubic decimeter. It is called a liter (litre). The same prefixes are applicable with respect to the liter.

The common temperature scale is called degree Celsius after the Swedish scientist, Anders Celsius, who created it.

SI unit names are spelled differently in different languages but SI symbols are the same in all languages and therefore should be shown the same everywhere. SI symbols for units and prefixes

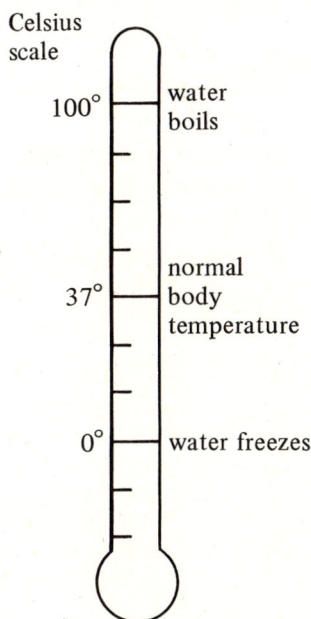

FIGURE 7.7

253

are not abbreviations, so they must be in the prescribed lettercase; thus m, not M, is the symbol for meter. SI symbols are not followed by a period or a plural "s". Table 7.2 lists some metric units and corresponding symbols.

TABLE 7.2   Symbols for Metric Units of Measurement

Type	Unit	Symbol
length	millimeter	mm
	centimeter	cm
	meter	m
	kilometer	km
area	square millimeter	mm^2
	square centimeter	cm^2
	square meter	m^2
	square kilometer	km^2
volume	cubic millimeter	mm^3
	cubic centimeter	cm^3
	cubic meter	m^3
	liter (cubic decimeter)	L
weight (mass)	gram	g
	kilogram	kg
temperature	degree Celsius	$°C$

You will need a metric ruler (15 or 30 centimeters long) for the work in this and later sections.

PROBLEM 1   a. Using your metric ruler determine the length of each to the nearest centimeter.
Your pen _____ cm
Your notebook _____ cm
Your shoe _____ cm
Your thumbnail _____ cm

b. Measure each line segment to the nearest millimeter. Record your measurement in the blank at the right.

_____   _____ mm

_____   _____ mm

_____   _____ mm

_____   _____ mm

c. Are the measurements you found in **a** and **b** exact? Why?

d. When measuring to the nearest centimeter, what is the greatest possible error you would make?

HOMEWORK EXERCISES

1. Find some convenient reference measures in metric units by measuring parts of your body such as hand span, width of a finger, distance between outstretched thumb and first finger. Using these reference measures only, draw line segments of the following lengths.

    a. 10 cm
    b. 2 cm
    c. 30 mm
    d. 70 mm
    e. 20 cm
    f. 15 cm
    g. 100 mm
    h. 100 cm
    i. 2 m
    j. 75 cm

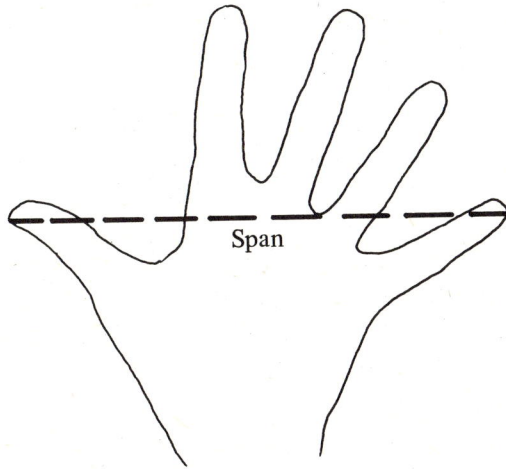

Span

FIGURE 7.8

Check each with a metric ruler. How close were you in each case? Try again if you need more practice. Since your eyes may see a length differently from another perspective, draw the first seven segments on paper. Were your drawings closer to the lengths prescribed?

2. Measure your pace, your height, and your arm span (distance between outstretched arms) in metric units. Using these reference measures and any from exercise 1 estimate:

    a. the length, width and height of a full size car, a compact car, and a subcompact;

    b. the diameter of a tire from each car.

    Check each estimate with a metric tape. How close were you in each case?

7.5   PERIMETER AND AREA OF POLYGONS

The concepts of polygon, triangle, quadrilateral, parallelogram, rectangle, and square were introduced in the previous chapter on congruence. You will now look at some specific types of polygons and investigate the relationship of their areas to each other and the derivation of area formulas you have probably seen before.

You will need some construction paper, your metric ruler, a right angle, and a scissors for some of the work in this and some of the later sections of this chapter.

The basis for the development of area formulas is the fact that the area measure of a rectangular region is calculated by multiplying the linear measures of two adjacent sides, often called the length and width.

PROBLEM 1
  a. On construction paper: draw a rectangle 7 cm by 5 cm. Label the longest dimension *l* and the adjacent side *w*.

  b. Draw a parallelogram congruent to the parallelogram *ABCD*, shown as Figure 7.9. Label the vertices *inside* the parallelogram as indicated.

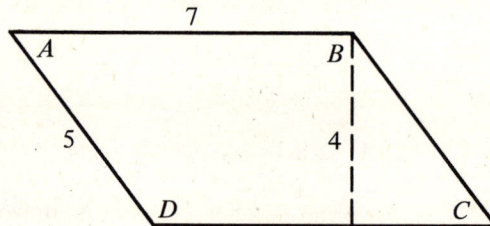

FIGURE 7.9

  c. What is the perimeter measure of the rectangle? What is the perimeter measure of the parallelogram *ABCD*?

  d. Draw 2 triangles like Figure 7.10 that are congruent to △*PQR*. Label *inside* the triangle. Label one *PQR* and the other *STV*.

  e. Draw a *right* triangle *XYZ* with legs 7 cm and 5 cm in length. (Legs

256

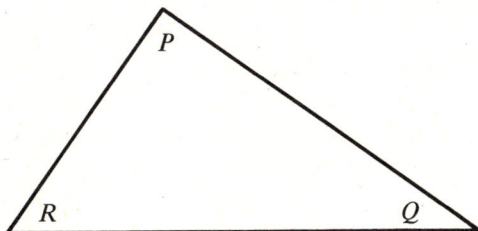

FIGURE 7.10

of a right △ are the sides forming the right angle.) Label it inside the triangle.

f. Cut out the rectangle, parallelogram, and triangles.

PROBLEM 2 **a.** Find the area measure of the rectangular region.

$$A_{rect} = l \cdot w = \underline{\hspace{2cm}} cm^2$$

b. What are the dimensions of the parallelogram?

c. Compare the cut-out rectangular region to the cut-out parallelogram region.
Which has the larger area measure if any?

d. How could you find the area measure of the parallelogram region?

e. Find the segment which contains B and is perpendicular to $\overline{CD}$. Label the point where it intersects $\overline{CD}$ as E. Draw $\overline{BE}$ on the parallelogram region and then cut the parallelogram region along the segment $\overline{BE}$. Rearrange the pieces into a rectangle. What are the dimensions of the rectangular region formed? What is the area? Generalize your findings and write a formula for finding the area of a parallelogram region.

PROBLEM 3 **a.** How can you find the measure of a triangular region?

b. Arrange triangles PQR and STV, side by side, to form a parallelogram region. Justify your claim that the figure is a parallelogram. How does the area measure of one of the triangular regions compare to the area measure of the parallelogram region?
How would you calculate the area measure of the parallelogram region? Write a formula for finding the area measure of a triangular region based on your observations.

c. Find the area of triangular region XYZ. Compare this cut-out triangular region to the cut-out rectangular region.

HOMEWORK EXERCISES

1. Find the perimeters and area measure, if possible, for the regions shown in Figure 7.11. If not possible, explain why not. Assume all the measures are given in cm.

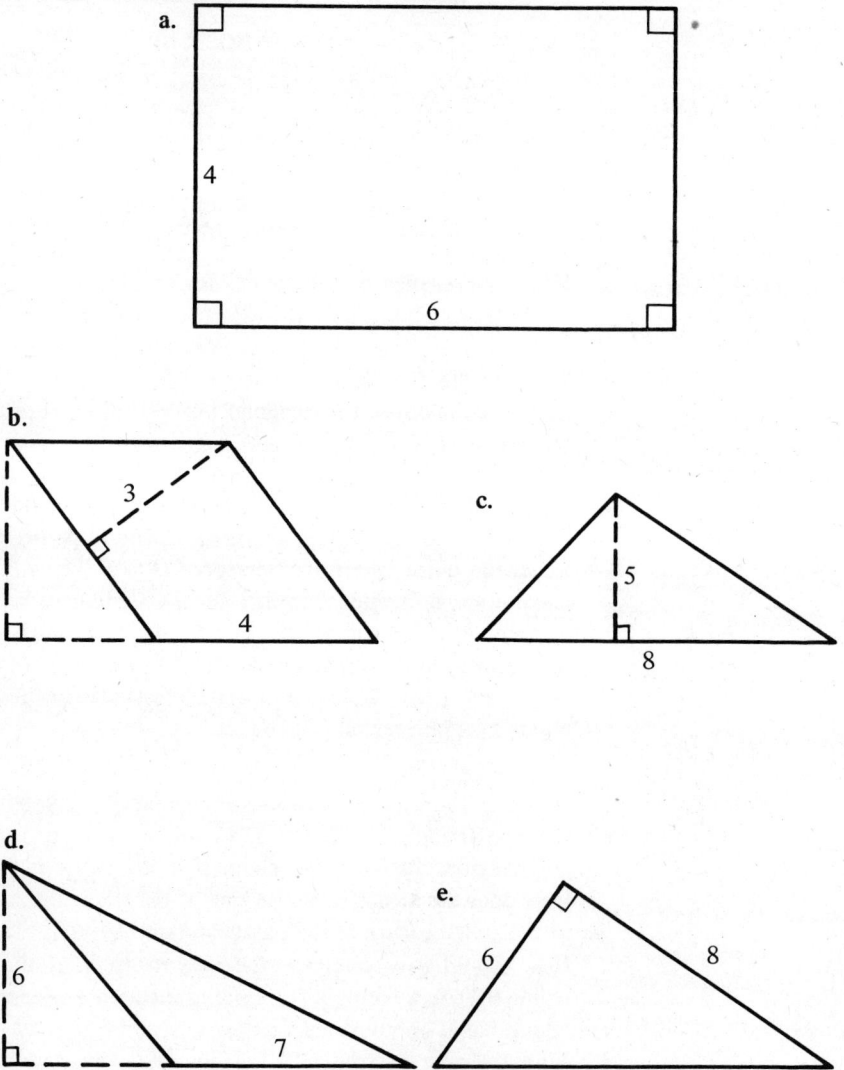

a.

4

6

b.

3

4

c.

5

8

d.

6

7

e.

6

8

FIGURE 7.11

2. A *trapezoid* is a quadrilateral with one pair of opposite sides parallel. Polygon *ABCD,* shown as Figure 7.12, is a trapezoid. Derive a formula for finding the area of a trapezoidal region.

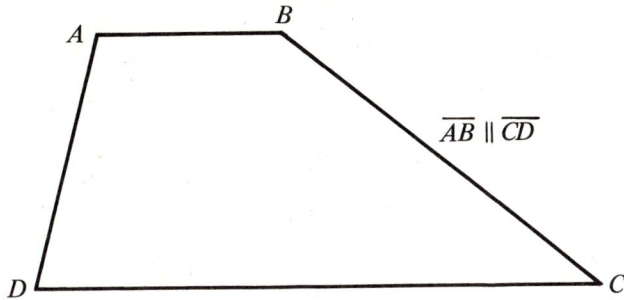

$$\overline{AB} \parallel \overline{CD}$$

FIGURE 7.12

3. Use the additive (subtractive) property of measure, if necessary, and the formulas previously developed to find the perimeters and area measures of the shaded regions in Figure 7.13. Measure those dimensions that you need to the nearest millimeter.

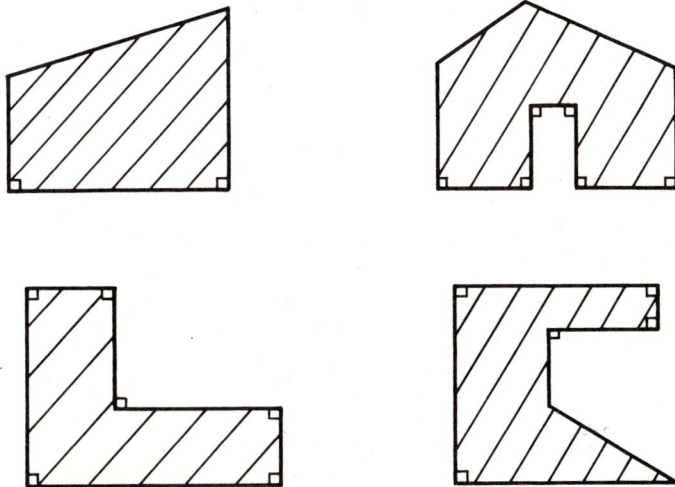

FIGURE 7.13

★4. A rhombus is a parallelogram with all sides congruent. You have shown in

Chapter 6 that the diagonals of a rhombus are perpendicular. Use this fact to derive a formula for the area of a rhomboidal region that does not depend on the linear measures of a side.

★5. Suppose you want to fence in a rectangular portion of your yard as a flower garden. You have 40 m of fencing.What shape would the flower garden be in order to maximize the area measure?

7.6 A SPECIAL PROPERTY OF A RIGHT TRIANGLE

PROBLEM 1   As accurately as possible, draw a right triangle on a sheet of construction paper, measuring at least 20-cm by 28-cm, with legs measuring 6 and 8 cm. Position your triangle so: **1.** the 8-cm side is parallel to and 8 cm from the bottom of the sheet, and **2.** the 6-cm side is parallel to and 6 cm from the left edge of the paper. Check to see that the measure of the hypotenuse (side opposite the right angle) is 10 cm. Refer to Figure 7.14.

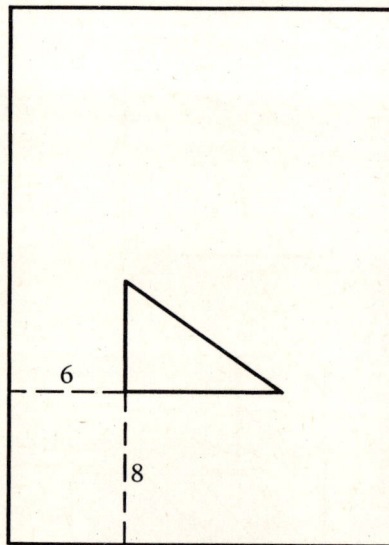

FIGURE 7.14

a. Construct squares on each side of the triangle as shown in Figure 7.15.

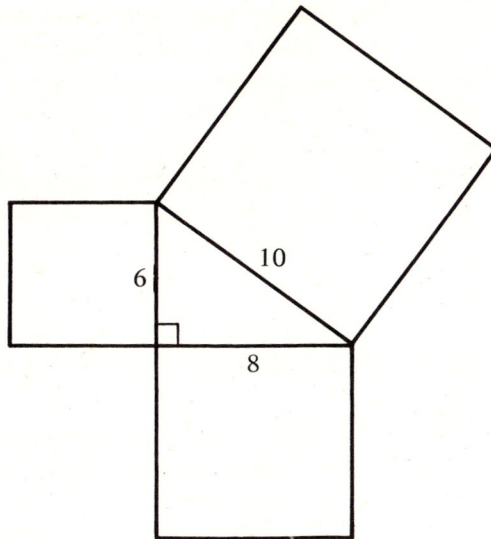

FIGURE 7.15

b. Carefully cut out the 8-cm square and the 6-cm square regions.
c. Carefully cut the 6-cm square region into two rectangular regions 2-cm by 6-cm and 3 square regions 2-cm by 2-cm. Refer to Figure 7.16.

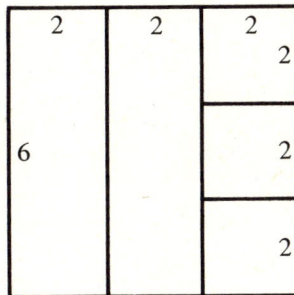

FIGURE 7.16

d. Now try to rearrange the 8-cm by 8-cm square region, the two rectangular regions, and the three 2-cm by 2-cm square regions so that they exactly cover the 10-cm by 10-cm square. Did you do it?

261

PROBLEM 2   a.   Show algebraically (numerically) why the exact covering of a 10-cm by 10-cm square region is possible using an 8-cm square region and a 6-cm square region.

b.   Do you think this relationship is true for all triangles? All right triangles? Why?

c.   State the algebraic relationship you found for a right triangle with legs whose measures are *a* and *b* and whose hypotenuse measure is *c.*

The example in the previous activity illustrates a very famous mathematical theorem known as the Pythagorean Theorem. It is named for the Greek mathematician Pythagoras, who discovered it. The fact that the relationship holds for *all right triangles* can be easily demonstrated.

PROBLEM 3   In Figure 7.17, let *a, b,* and *c* denote the measures of the legs and hypotenuse of a right triangle respectively. We wish to prove: $a^2 + b^2 = c^2$. We can construct a square whose sides measure $a + b$ containing four congruent copies of the given *C* as indicated in Figure 7.18.

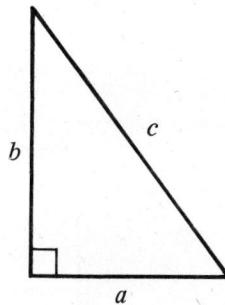

FIGURE 7.17

a.   Show that angles *HBD, BDF, DFH, FHB* are all right angles.
b.   What kind of figure is *HBDF*?
c.   Find the area of square region *ACEG* using the measure $a + b$. Express the area measure in terms of *a* and *b*.
d.   Now find the area of square region *ACEG* applying the additive property of measure to the subregions formed by adding $\overline{HB}, \overline{BD}, \overline{DF},$ and $\overline{FH}$ to the figure.

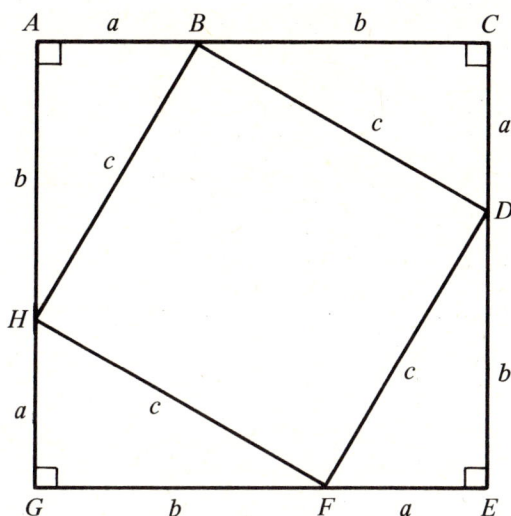

FIGURE 7.18

e. Since the area measure of square region *ACEG* should be the same no matter how computed, the two results should be equal. Set them equal to each other and algebraically simplify the equation. Have you given the desired result? Were there any restrictions on the original triangle?

The Pythagorean Theorem can be used to find the measure of any one side of a right triangle if the measures of the other two sides are known.

PROBLEM 4 Suppose a right angle has legs whose measures are 3 cm and 4 cm.
   a. Construct a congruent copy of the triangle.
   b. If *c* denotes the measure of the hypotenuse of the triangle, write the algebraic relationship of 3, 4, and *c* implied by the Pythagorean Theorem.
   c. Find the value of *c* by solving for *c* in the equation you wrote.
   d. Measure the length of the hypotenuse of the triangle you constructed. What is the measure?_____ Does this agree with the results obtained by applying the Pythagorean Theorem?
   e. Measure the sides of the right triangle shown as Figure 7.19. Check to see if the measures found conform to the Pythagorean Theorem. Did they? If not, why not?

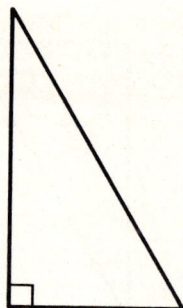

FIGURE 7.19

HOMEWORK EXERCISES

1. In Figure 7.20, if x denotes the measure of the side in the triangles that is not given, find the value of x, if possible. If not possible, explain why not.

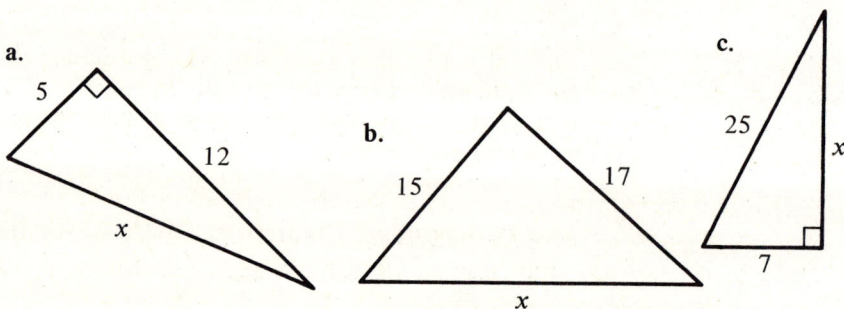

FIGURE 7.20

2. Triples A, B, C of *counting numbers* that have the Pythagorean relationship, $a^2 + b^2 = c^2$, are called Pythagorean Triples. For example: 3, 4, 5 is a Pythagorean Triple. Find several other Pythagorean Triples. Is there a technique for generating more Pythagorean Triples if you know one triple?

3. Rectangle A is x centimeters long and y centimeters wide.
   a. Rectangle B is 3 times as long and 3 times as wide as rectangle A. How do their areas compare?
   b. A parallelogram has the same base length and height as rectangle A. How do their areas compare?

   c. A parallelogram has the same height as rectangle *A* but the base length is twice as large. How do their areas compare?

   d. A triangle has the same base length but the height is twice the height of rectangle *A*. How do their areas compare?

4. Find the area of the trapezoid shown as Figure 7.21.

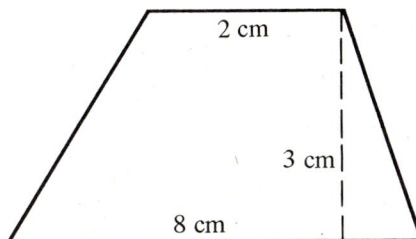

FIGURE 7.21

5. The hypotenuse of a right triangle is $\sqrt{26}$ in centimeters. What are the lengths of the other two sides?

7.7 REAL NUMBERS

The Pythagorean Theorem holds for all right triangles. However, not all right triangles have sides with whole number measures. We can certainly construct a right triangle with both legs 1 unit in length. According to the Pythagorean Theorem, the measure of the hypotenuse, *c*, would satisfy the relationship $c^2 = 1^2 + 1^2 = 1 + 1 = 2$. Thus *c* is the number whose square is 2 or equivalently *c* equals the *square root* of 2 denoted $\sqrt{2}$.

PROBLEM 1 Choose a unit and, in the space below, construct a number line that runs 0 units to 4 units.

   a. Indicate the location of the number, $\sqrt{2}$, as precisely as you can, on your number line. Justify your choice of location.

   b. Suppose $\sqrt{2}$ could be represented as a fraction, $\frac{a}{b}$, where a, $b \in Z$ and $b \neq 0$. What kind of number would $\sqrt{2}$ be if this were true?

   c. Could you locate $\sqrt{2}$ more accurately on your number line if you knew a fraction equivalent to $\sqrt{2}$? If so, how?

You can investigate whether or not there exists a fraction that represents $\sqrt{2}$ without knowing what it is if it *does* exist. Some of the results you saw in Chapter 4 concerning divisors and primes will be put to use as well as some other number theoretic relationships not previously used.

★PROBLEM 2　Choose numerical examples to illustrate the following statements and to convince yourself that they are true.

Let p be a prime number and a an integer.

STATEMENT 1:　If $p \mid a^2$ then $p \mid a$

STATEMENT 2:　If $p \mid a$ then $p^2 \mid a^2$

Suppose there is a fraction, $\frac{a}{b}$, where a, $b \in Z$ and $b \neq 0$, that represents $\sqrt{2}$. You can assume the fraction is in lowest terms; that is, there is no positive integer other than 1 that divides both a and b. (If $\frac{a}{b}$ is not in lowest terms, you could find an equivalent fraction that is!)

So
$$\frac{a}{b} = \sqrt{2}$$

then
$$\left(\frac{a}{b}\right)^2 = (\sqrt{2})^2$$

or
$$\frac{a^2}{b^2} = 2$$

and
$$a^2 = 2b^2$$

now
$$2 \mid 2b^2, \text{ so } 2 \mid a^2$$

By Statement 1:

$$2 \mid a^2, \text{ therefore } 2 \mid a$$

By Statement 2:

$$2 \mid a, \text{ so } 2^2 \mid a^2 \text{ or } 4 \mid a^2$$

Since $4 \mid a^2$, $4 \mid 2b^2$

From the definition of *divides* (Chapter 4)

$$4k = 2b^2$$
and
$$2k = b^2$$
$$2 \mid 2k \text{ so } 2 \mid b^2$$

Using Statement 1 again

$$2 \mid b^2, \text{ so } 2 \mid b$$

Now you have $2 \mid a$ and $2 \mid b$ which means $\dfrac{a}{b}$ is not in lowest terms.

This is a contradiction! So your assumption that there is a fraction which represents $\sqrt{2}$ is false. What this says, then, is that the number, $\sqrt{2}$, is not a rational number.

There are many other numbers that are not rational, like $\sqrt{3}$, $\sqrt{5}$, and π (a number you will investigate later). The set of rational numbers together with the set of *irrational* numbers forms the set of *real numbers* represented by the points of the number line.

Since you are not attempting a formal study of the real number system, you will consider them in terms of their decimal representations. Burton Jones, in his book *Elementary Concepts of Mathematics,* suggests that the simplest thing to do is to take advantage of the inventive genius of the first person who said, "Let there be enough numbers so that there is one for each nonterminating or terminating decimal, and let these numbers be called *real numbers.*"

Since the set of real numbers includes the set of rational numbers as well as the set of irrational numbers, look at the decimal representations of both sets.

You recall from Section 2.9 that the set of rational numbers includes all quotients of integers with denominators *never* zero. Thus rational numbers can be named by fractions like $\dfrac{1}{4}$, $\dfrac{^-5}{6}$, $\dfrac{6}{1}$.

Decimal fractions (or simply decimals) are used to represent irrational numbers.

PROBLEM 3 **a.** Can rational numbers be represented by decimals? If so, how can you find the decimal equivalent for any fraction?

b. Find decimal representations for the rational numbers represented by $\dfrac{2}{5}$, $\dfrac{4}{50}$, $\dfrac{1}{3}$, and $\dfrac{2}{7}$.

c. Did you encounter any difficulty finding the decimal equivalent to the fractions in part **b**? Did you find exact equivalents? Explain.

An easy way to find the decimal equivalent to the fraction $\dfrac{a}{b}$ is to divide the denominator, *b*, into the numerator, *a*. The resulting quotient will eventually repeat zeros or a block of one or more digits. Use an overbar to indicate the block of digits that repeats. It is understood that repeating zeros are dropped. Decimals of this type are called *repeating decimals*.

Examples

$$\frac{2}{5} = .400\overline{0} = .4$$

$$\frac{1}{3} = .33\overline{3}$$

$$\frac{2}{7} = .\overline{285714}$$

The following statement establishes the relationship of rational numbers (and their fractions) to repeating decimals. *Every rational number can be expressed as a repeating decimal, and every repeating decimal represents a rational number.*

PROBLEM 4 Since every repeating decimal represents a rational number, there must be a fraction equivalent to each repeating decimal.

a. Find fractions equivalent to each of the repeating decimals.

$.66\overline{6} =$

$1.5\overline{2} =$

.68 =

.22$\overline{2}$ =

Check your work by changing your fractions back to decimals. You may use a calculator to help if one is available.

Did you get $\dfrac{2}{9}$ for the last one? You should have!

b. Converting a repeating decimal to an equivalent fraction is not a trivial task when faced with decimals like .$\overline{142857}$ $\left(\dfrac{1}{7}\right)$ or .1632$\overline{32}$ $\left(\dfrac{1616}{9900}\right)$.

Consider the following procedure if 1 digit repeats:

Step	$x = .3211\overline{1}$
1. Mult. by <u>10</u>	$10x = 3.211\overline{1}$
2. Subtract x from $10x$	$\begin{aligned} 10x &= 3.211\overline{1} \\ x &= .321\overline{1} \\ \hline 9x &= 2.89 \end{aligned}$
3. Multiply both sides by <u>100</u>	$900x = 289$
4. Solve for x	$x = \dfrac{289}{900}$

If 2 digits repeat:

Step

$y = .48\overline{48}$

1. $\begin{aligned} 100y &= 48.48\overline{48} \\ y &= .48\overline{48} \\ \hline \end{aligned}$

2. $99y = 48$

4. $y = \dfrac{48}{99}$

c. The two examples in part **b** are given to suggest a more general procedure. The numbers underlined in each step vary according to the decimal being converted. Step 3 is only necessary sometimes. Try to generalize the procedure so it will work for any repeating decimal. Construct a flowchart for your procedure and test it on the repeating decimals below.

.77$\overline{7}$ =

.6038$\overline{38}$ =

2.514$\overline{514}$ =

.4$\overline{89}$ =

$$1.99\overline{9} =$$
$$.46152 =$$

d. Did any of the repeating decimals in part **c** cause particular difficulty? If so, which ones and why?

e. Note that the way decimals are written allows more than one decimal name for some numbers.

$.49\overline{9} = .5$

$^-.99\overline{9} = ^-1.0$

$.58 = 579\overline{9}$

Find a second decimal name for each number below.

$6.0 =$

$1.5 =$

$3.\overline{9} =$

$.32 =$

$1.813 =$

$.137\overline{9} =$

PROBLEM 5 What does the statement relating rational numbers to repeating decimals imply about the representation of irrational numbers?

The decimals representing irrational numbers are infinite decimals that are not repeating. They can never be written in their entirety. So, one must decide how much accuracy is necessary for the task at hand and approximate the irrational number with a rational approximation. The decimal .1011011101111 ... is an example of an infinite non-repeating decimal. It has a pattern but does not infinitely repeat a block of digits. To use the number represented by .1011011101111 ... in a calculation we would have to round-off or truncate to some appropriate number.

Another technique which is very useful in dealing with certain irrational numbers is to apply the following rules of square roots to simplify square root expressions. $a \geqslant 0, b > 0$

$$\sqrt{a} \cdot \sqrt{b} = \sqrt{ab}$$

$$\frac{\sqrt{a}}{\sqrt{b}} = \sqrt{\frac{a}{b}}$$

Thus,
$$\sqrt{3} \cdot \sqrt{12} = \sqrt{36} = 6$$
$$\sqrt{24} = \sqrt{4 \cdot 6} = \sqrt{4} \cdot \sqrt{6} = 2\sqrt{6}$$

These rules will be especially useful in the next chapter.

HOMEWORK EXERCISES

1. Write each fraction as a repeating decimal.
   a. $\dfrac{11}{7}$

   b. $\dfrac{2}{3}$

   c. $\dfrac{3}{5}$

   d. $\dfrac{7}{25}$

   e. $\dfrac{1}{16}$

   f. $\dfrac{3}{4}$

2. Express each repeating decimal as a fraction.
   a. .0625
   b. .125
   c. $.142\overline{142}$
   d. $1.99\overline{9}$
   e. $.83\overline{83}$

3. Which of the following real numbers are irrational?
   a. $^-\sqrt{25}$
   b. $2\sqrt{5}$
   c. 3.14
   d. 2π
   e. $\dfrac{2}{3}$

4. Simplify each of the following expressions.
   a. $\sqrt{48}$

   b. $\sqrt{\dfrac{9}{4}}$

    c. $\sqrt{8} \cdot \sqrt{7}$
    d. $\sqrt{.5}$
    e. $\sqrt{.25}$

5. $\sqrt{5}$ is irrational. May a segment be $\sqrt{5}$ units in length? Explain.

★6. When converting the fraction $\dfrac{a}{b}$ into an equivalent decimal by dividing b into a, you know that eventually the digits of the decimal will begin to repeat in blocks. What is the maximum number of division steps that must be carried out in order to have the repetition begin? **Hint:** Consider the role of remainders when performing the division process.

★7. Show that the product of a nonzero rational number and an irrational number is irrational. **Hint:** Use proof by contradiction.

★8. You recall that the set of rational numbers, with the operations addition and multiplication, is an example of a field. The real numbers also form a field under addition and multiplication. The irrational numbers, however, do not satisfy all the field properties. Illustrate which of the field properties fail in the set of irrational numbers.

7.8 CIRCLES AND CIRCULAR REGIONS

PROBLEM 1   **a.** The Greek letter π is the irrational number that is the ratio of the length of a circle to its diameter. The number line shown as Figure 7.22 is scaled in units the same length as the diameter of a dime. Mark some starting point on a dime (the bottom of the 1 in the date, for example) and placing the marked point on zero, roll the coin on the line one full revolution. Mark the end point of the revolution on the number line. Be careful to minimize slipping.

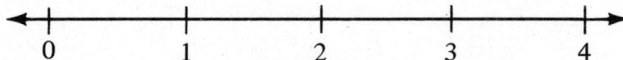

FIGURE 7.22

The distance from 0 to your marked point on the line is the length of the circle (circumference). What length is it?_____

b. Even though π is irrational, you are probably familiar with one or two rational approximations for π.

$$\pi = 3.141592\ldots$$

List the two common rational approximations used for π and show that neither of them is equal to π. Which is closer to the actual value of π?

c. Since π is defined as the ratio of the length of a circle, c, to the diameter, d, then

$$\pi = \frac{c}{d} \text{ or } c = \pi d$$

When finding the length of a circle, you can either express the length in terms of π (exact) or use an approximation for π and thus obtain an approximate value.

For each of the following problems, express the result as an exact solution and also as an approximation.

1. Find c when $d = 4$
2. Find d when $c = 27$

PROBLEM 2   On construction paper, carefully draw (use a compass) 3 circles with 15-cm diameters.

a. Carefully divide and cut one circular region into four congruent sectors as shown in Figure 7.23.

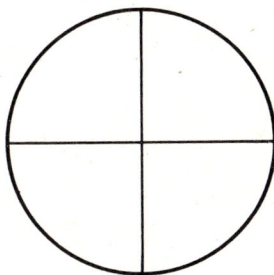

FIGURE 7.23

b. As carefully as you can, divide the second circular region into six congruent sectors and the third into twelve congruent sectors as shown in Figure 7.24.

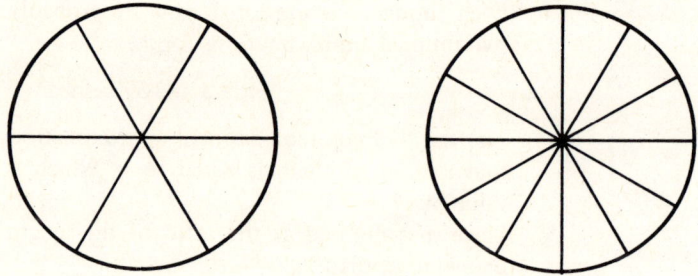

FIGURE 7.24

PROBLEM 3 **a.** Rearrange the sectors of the first circular region as indicated in Figure 7.25.

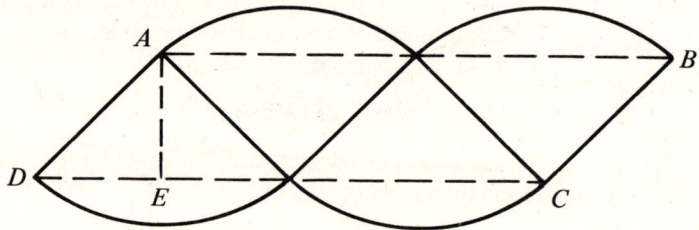

FIGURE 7.25

Construct $\overline{AE}$ perpendicular to $\overline{DC}$.
Measure $\overline{DC}$ and $\overline{AE}$ (nearest mm).

$m(\overline{DC}) = $ _____ mm

$m(\overline{AE}) = $ _____ mm

Approximate the area of the circular region by finding the area of the parallelogram region $ABCD$.

Approximate area of circular region = _____ mm². Is your approximation too large or too small? Explain.

b. Arrange the sectors of the second circular region as indicated in Figure 7.26. Repeat the approximation procedure.

$m(\overline{DC}) = $ _____ mm

$m(\overline{AE}) = $ _____ mm

Approximate area of circular region = _____ mm².

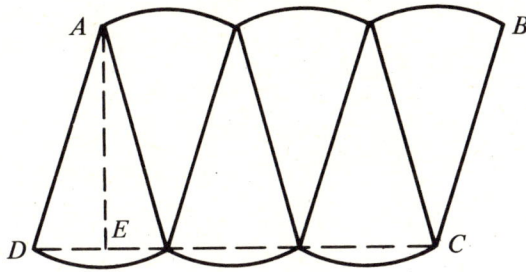

FIGURE 7.26

Is this approximation too big or too small?

Compare this second approximation to the first.

Which is closer to the actual area of the circular region? Why?

Compare the measures of $\overline{DC}$ and $\overline{AE}$ for the two approximations. What do you notice?

c. Do the same thing again for the third circular region.

   $m(\overline{DC}) = $ _____ mm

   $m(\overline{AE}) = $ _____ mm

   Approximate area of circular region = _____ mm^2. Compare the measures of $\overline{DC}$, $\overline{AE}$, and areas for the three approximations and write a generalization of what is happening.

d. If this procedure is continued indefinitely, the measure of $\overline{DC}$ approaches ½ the measure of the _____ of the circle and the measure of $\overline{AE}$ approaches the measure of the _____ of the circle and the area of the parallelogram region approaches the area of the circle.

   Use these conclusions to write a formula for the area of a circular region.

   Area of a circular region = ...

HOMEWORK EXERCISES

1. A *cylinder* is a solid region with circular regions for bases. Find the surface area of the cylinder in Figure 7.27.

2. Find the area of circular region bordered by a circle whose length is

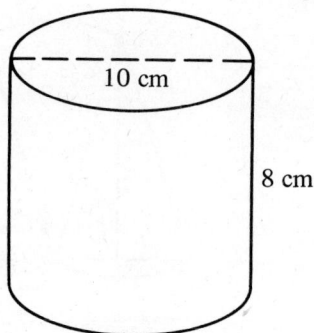

FIGURE 7.27

40 m. Compare your result to the result of Homework Exercise 5 in Section 7.5. What does this suggest?

3. Find the area of the shaded portion of Figure 7.28.

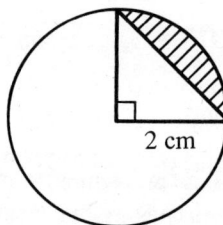

FIGURE 7.28

4. A dog is tied at the corner of a rectangular house with dimensions 150 cm and 75 cm. The rope is 100 cm long. Ignoring the inside of the dog house, what is the measure of the area the dog can roam?

5. A circular rug two meters in diameter costs $96. A shopper figured that a similar rug four meters in diameter should cost twice as much. How would you explain the shopper's error? If size alone determined the price, the larger rug will cost _____ times as much as the smaller one.

★6. Show that a circular region bordered by a circle with length P units has greater area than any rectangle with perimeter P units.

7.9   VOLUME

You have looked at linear (length) measure of segments and area measure of plane regions. The volume of solid regions is your next concern. Volume measure is a measure of the amount of space occupied by a 3-dimensional (solid) region. You will restrict your development to the common geometric solids: prisms, cylinders, pyramids, and cones. Figure 7.29 illustrates some of these solid regions.

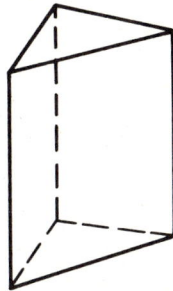

Triangular prism   Rectangular prism   Cylinder

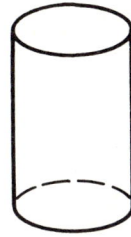

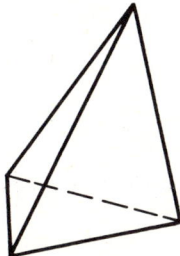

Triangular pyramid   Rectangular pyramid   Cone

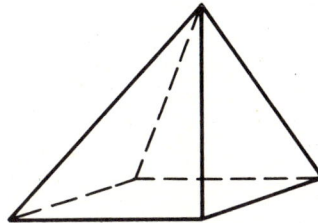

FIGURE 7.29

PROBLEM 1   Write definitions for each of the geometric solids shown in Figure 7.29.

PROBLEM 2   Using the fact that the volume of a rectangular prism can be found by multiplying the length measures of its length, width, and height, derive a formula for finding the volume of a triangular prism.

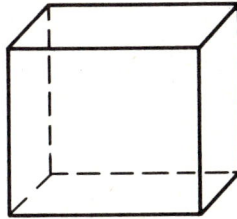

A general formula for finding the volume of either a prism or cylinder is $V = Bh$, where B represents the area measure of the base and h is the linear measure of the height of the figure.

PROBLEM 3   **a.** Compare the results you obtained for Problem 2 with the general formula. Are the formulas compatible?

            b. Show that the volume formula for a rectangular prism, $V = lwh$, fits the generalized formula.

HOMEWORK EXERCISES

1. Think of a 3-centimeter cube with the entire surface painted yellow. (You may want to sketch a picture of this cube.)
   a. Into how many 1-centimeter cubes could you cut this cube?
   b. How many cuts are required to divide it into 1-centimeter cubes?
   c. How many 1-cm cubes will have 4 yellow faces?_____ 3 yellow faces?_____ 2 yellow faces?_____ 1 yellow face?_____ 0 yellow faces?_____
   d. How many square centimeters of surface area are on this 3-cm cube?

      _____

   e. The volume of this 3-cm cube is _____ .

2. A room is 6 meters long, 5½ meters wide and 3 meters high.
   a. Find the total area of the walls and ceiling allowing 7 square meters for windows and doorway.
   b. How many liters of paint will be needed to paint the walls and ceiling if a liter covers about 8½ square meters?

3. A schoolroom measures 10 by 8 by 4 meters. How many cubic meters of air space are there for each student if there are 30 students in the room?

   _____

4. a. An excavation has the following dimensions: 2 meters deep, 7 meters wide, and 12 meters long. How many cubic meters of dirt were removed? _____
   b. A tank 30 cm by 50 cm by 20 cm has a volume or capacity of_____. You may want to know how much water it will hold. A liter is the basic unit of volume or capacity. It is defined in terms of the meter. Since the cubic meter is too large and the cubic centimeter is too small, a more reasonable basic measure for volume (capacity), the

cubic decimeter, was chosen. One cubic decimeter is 1000 cubic centimeters. The measure is called the *liter*. (A 10 by 10 by 10 Dienes block has a volume of 1000 cubic centimeters. A half gallon milk carton cut off 10 cm high is also a good representation of this basic unit which is called a liter.) How many liters of water will fill the tank in this problem?

Suppose you consider a cone and cylinder with the same base and height. Is there a relationship between the volumes of the two solids? The next activity and problem will investigate this relationship.

You will need a supply of a dry pourable substance (salt, sugar, sand, kitty litter, etc.), some tape, and scissors for the next activity.

PROBLEM 4 a. Cut out congruent copies of the regions shown as Figure 7.30. Assemble them so that you obtain a cylinder (open on one end) and a cone (open base). Check that the bases and heights of the solids are the same measure.

b. Fill the cone to the brim with the dry pourable substance and then pour it into the cylinder. Repeat the procedure until the cylinder is full. Keep track of the number of times you fill the cone.

PROBLEM 5 a. Write a formula for finding the volume of a cone. Explain any variables used in the formula.

b. Do you think the relationship between the volume of a pyramid and prism with congruent base and height is the same as that for a cone and cylinder? How could you check?

HOMEWORK EXERCISES

1. Find the volume for each solid region.
   a. A right-triangular prism with legs of 4 cm and 5 cm and height of 10 cm.
   b. A rectangular prism with dimensions 7 cm, 8 cm, and 9 cm.
   c. A square pyramid with an edge of the base measuring 2 cm and a height of 4 cm.
   d. A cylinder with a base whose perimeter is equal to that of the triangular prism in part **a** and a height of 10 cm.

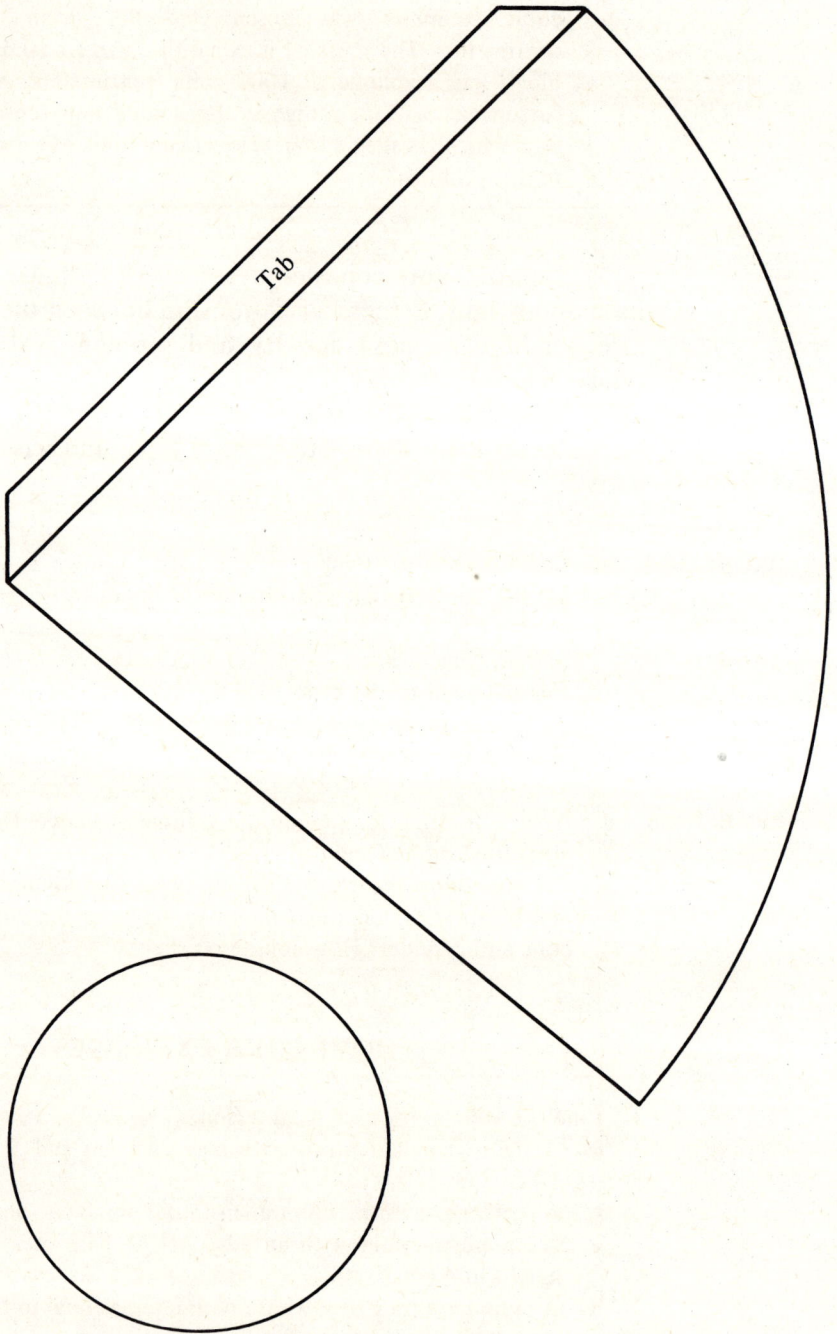

Tab

FIGURE 7.30

Tab

FIGURE 7.30 (cont.)

2. Find the volume of a cone whose base has a perimeter of 30π cm and a height of 6 cm.

★3. The volume of a sphere can be obtained by inscribing a cone and circumscribing a cylinder in and about a hemisphere, taking the average of their volumes and doubling that number. In terms of r, the radius of the sphere, what is its volume?

SUMMARY

Measurement is a very basic idea—used by you even before you went to school. This simple mapping which assigns a number to a characteristic of an object enables you to organize the world. Measure is a mapping that structures and orders the objects around you. Very little material progress in the civilized world is possible without measurement. All of the sciences, social, biological, and physical, depend on measurement. The characteristics are varied. Length, area, volume, surface area, circumference, weight, angle measure, and temperature were the ones used in Chapter 7. Investigation of the measure of the hypotenuse of many right triangles led you, as it did the early Greeks, to an extension of the number systems to include the real numbers. Now the counting numbers, the integers, the rational numbers, and the real numbers are available sets. Other sets of numbers are needed in mathematics, but they will not be considered in this text. Of course, while examining measurement, the processes of mathematics were utilized. A particularly important one in Chapter 7 is the application of the mathematical model of measure to the physical world. The model is a good one; it fits the physical world rather well. However, in the physical world measurements will seldom be exact even though they are exact in the model.

Similarity

8

8.1 ANOTHER MAPPING

PROBLEM 1   **a.** Using the pair of coordinate axes in Figure 8.1, with O as the origin, find the images of A, B, and C under the mapping with the rule $(x,y) \rightarrow (kx,ky)$ where $k = 2$. Label these image points A', B', C' respectively.

      b. Draw $\overrightarrow{OB}$, $\overrightarrow{OA}$ and $\overrightarrow{OC}$. What do you notice about points A', B', C'?

      c. Using the Pythagorean Theorem find:

$$m(\overline{OA'}) =$$
$$m(\overline{OA}) =$$
$$m(\overline{OB'}) =$$
$$m(\overline{OB}) =$$

Simplify your results as much as possible using the rules for manipulation of square roots, and compare the first distance to the second in each pair. What relationship did you find?

Check your conjecture with $m(\overline{OC'})$ and $m(\overline{OC})$.

      d. Draw $\triangle ABC$ and $\triangle A'B'C'$ in the diagram.

$$m(\overline{A'B'}) =$$
$$m(\overline{AB}) =$$

Compare the relationship of the measures of the other two sides of $\triangle ABC$ to their images.

Generalize your findings.

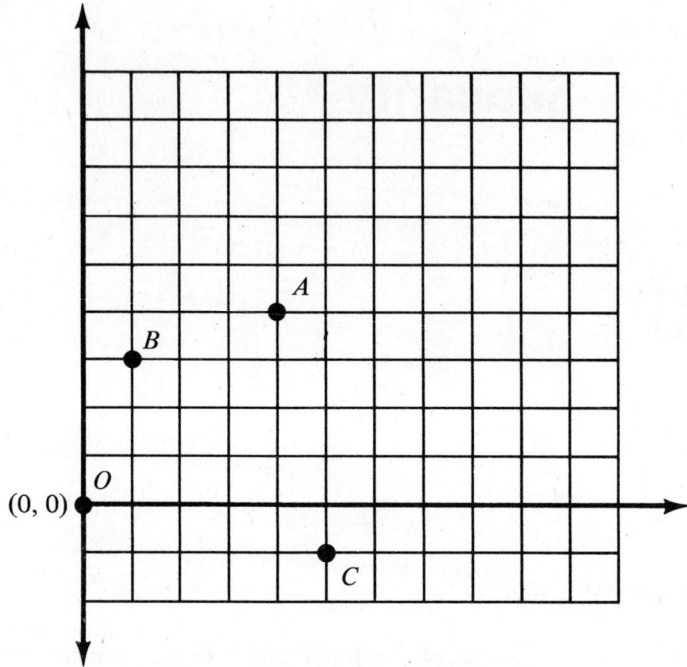

FIGURE 8.1

The mapping given in Problem 1 (Figure 8.1) is an example of a *dilation*. Dilations are mappings of the plane to the plane that can be thought of as "stretchers" and "shrinkers." A dilation has a *center* and a positive real number *scale factor* as defining characteristics. Each point of the plane has an image point for a specific dilation. In this regard, dilations are somewhat like the rigid motions we looked at in Chapter 6.

PROBLEM 2   Review the summary chart you completed for each type of rigid motion, Table 6.2, page 222. List the kinds of properties investigated and note any special properties that were true for all rigid motions.

The following set of homework exercises will focus your attention on some of the properties of dilations.

HOMEWORK EXERCISES

1. On $\overrightarrow{PS}$ find an image point that is 3 times as far from P as S is and label it S'. Find images of R and T on $\overrightarrow{PR}$ and $\overrightarrow{PT}$ in the same manner. Draw $\triangle RST$ and $\triangle R'S'T'$.

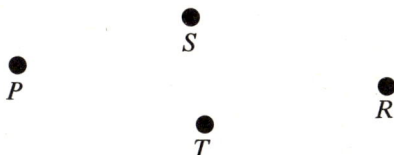

S

P

R

T

FIGURE 8.2

2. Using your ruler scaled in centimeters compare the measures of the sides of the triangles.

   $m(\overline{RS})$ =_____ cm
   $m(\overline{ST})$ =_____ cm
   $m(\overline{RT})$ =_____ cm
   $m(\overline{R'S'})$ =_____ cm
   $m(\overline{S'T'})$ =_____ cm
   $m(\overline{R'T'})$ =_____ cm

3. Using your protractor compare the measures of the angles of the two triangles.

   $m(\angle R)$ =_____
   $m(\angle S)$ =_____
   $m(\angle T)$ =_____
   $m(\angle R')$ =_____
   $m(\angle S')$ =_____
   $m(\angle T')$ =_____

4. Does it seem that: $\overline{RS} \parallel \overline{R'S'}$, $\overline{ST} \parallel \overline{S'T'}$ and $\overline{RT} \parallel \overline{R'T'}$?

5. In Figure 8.3, let point P, interior to $\triangle ABC$, be the center of the dilation.

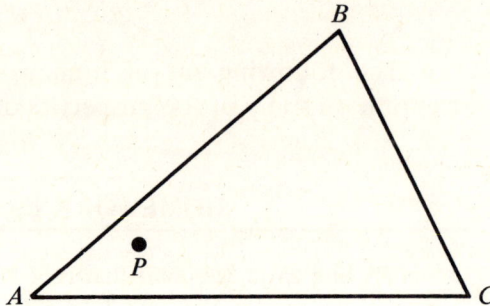

FIGURE 8.3

Repeat questions 1, 2, 3, 4 using P as the center of the dilation and ½ as the scale factor.

6. In the dilation in Exercise 5, $P \rightarrow$ _____ .

$A \rightarrow$ _____ on ray _____

$B \rightarrow$ _____ on ray _____

$C \rightarrow$ _____ on ray _____

Can you generalize for any center of dilation P and scale factor k how any point X will be mapped? What will be the measure of $\overline{PX'}$? Write the definition of such a dilation. (You will use the symbol $D_{P,k}$ to represent a dilation with center P and scale factor k.)

PROBLEM 3   Summarize the results of the previous Homework Exercises.

Let $D_{P,k}$ represent a dilation with center P and scale factor k.

a. If $X \overset{D_{P,k}}{\rightarrow} X'$, then $X' \in$ _____ .

b. $m(\overline{PX'}) =$ _____ $\cdot m(\overline{PX})$.

c. $P \overset{D_{P,k}}{\rightarrow}$ _____ .

Thus, $D_{P,k}$ is a mapping of the plane to the plane such that $P \rightarrow P$ and each $X \neq P$ maps to X' such that _____ .

PROBLEM 4   Since dilations are mappings of the plane to the plane, you can compare the properties of dilations to the properties of rigid motions.

    a. Investigate $D_{P,k}$ for points that map to themselves, lines that map to themselves, and lines that map to parallel lines.

    b. Do dilations map figures to figures of the same type? (Does Axiom 1 of Section 6.11 hold for dilations?)

    c. Do any figures map to congruent figures? Which ones?

    d. What characteristics, if any, of geometric figures are preserved by dilations?

HOMEWORK EXERCISES

7. Only using straight edge, compass, and pencil, locate the image of X for each dilation in Figure 8.4, and label the image point as indicated.

    a. $D_{P,\frac{1}{2}}(X')$

    b. $D_{Q,3}(X'')$

    c. $D_{Q,\frac{1}{4}}(X^*)$

    d. $D_{P,\frac{3}{2}}(X^{**})$

    e. $D_{Q,\frac{1}{4}} \circ D_{Q,3}(X\#)$

    f. $D_{R,\frac{1}{3}}(X\#\#)$

FIGURE 8.4

8. Consider a coordinate plane. Let P be the origin (coordinates $(0,0)$). Let A and B be points in the plane with coordinates $(1,4)$ and $(5,1)$ respectively.

    a. From Problem 1 of Section 7.1, $(x,y) \overset{D_{P,k}}{\rightarrow} (kx,ky)$. Find the coordinates of A' and B' if $A \overset{D_{P,3}}{\rightarrow} A'$ and $B \overset{D_{P,3}}{\rightarrow} B'$.

    b. In a coordinate plane, the slope of a line, or line segment, can be found by the following formula: $S = \dfrac{y_2 - y_1}{x_2 - x_1}$ where (x_1, y_1) and

(x_2, y_2) are the coordinates of *any* two points of the line or line segment. Find the slopes of $\overleftrightarrow{AB}$ and $\overleftrightarrow{A'B'}$.

★c. A definition of parallel lines could be stated in terms of slopes. Write the definition.

Use your definition and the fact that $(x, y) \overset{D_{P,k}}{\to} (kx, ky)$ when P has coordinates $(0,0)$ to prove that every line is parallel to its image line under a dilation.

8.2 A SYSTEM OF DILATIONS

Dilations are mappings, so consider what happens when you compose two dilations. Restrict your initial inquiry to the set of all dilations with center P, where P is any particular point. Let D_P represent this set of dilations; then (D_P, o) is the operational system formed by the composition of dilations in D_P.

PROBLEM 1 a. Is (D_P, o) closed? Is $D_{P,k} \circ D_{P,m}$ a dilation with center P? If so, and $D_{P,k} \circ D_{P,m} = D_{P,r}$, what is the relationship between k, m, and r? Check several examples.

b. Check the rest of the properties of (D_P, o). Show counter-examples, if appropriate.

PROBLEM 2 Suppose $D_{P,x} \circ D_{P,y} = D_{P,x \cdot y}$. Show how this property can be used to prove (D_P, o) is a commutative group.

★**PROBLEM 3** Let $\mathscr{D} = \{$all dilations of the plane$\}$. Investigate the structure of $(\mathscr{D}, o)$. Is $D_{P,x} \circ D_{Q,y}$ a dilation?

8.3 DEDUCTIONS BASED ON DILATIONS

Dilations and their properties can be used to prove some familiar geometric facts in much the same way that rigid motions and the properties of rigid motions were used. Work out one example to demonstrate.

PROBLEM 1   **a.** Locate the midpoints of $\overline{XZ}$ and $\overline{XY}$ of $\triangle XYZ$. Label the midpoints P and Q respectively. Draw $\overline{PQ}$.

b. Conjecture some relationships between $\overline{PQ}$ and $\overline{ZY}$.

c. Consider $D_{X,2}$.

By the definition of a dilation:

$$P \overset{D_{X,2}}{\to} \underline{\hspace{2cm}}$$

$$Q \overset{D_{X,2}}{\to} \underline{\hspace{2cm}}$$

thus $\overline{PQ} \overset{D_{X,2}}{\to} \overline{ZY}$. Why?

Since $\overline{ZY}$ is the image of $\overline{PQ}$ for a dilation with scale factor 2, what do you know about the relationship of $\overline{PQ}$ to $\overline{ZY}$?

d. State the relationship proved above in the form of a theorem.

HOMEWORK EXERCISE

★**1.** Suppose line m intersects sides $\overline{AB}$ and $\overline{AC}$ of $\triangle ABC$ in points E and F respectively. Show that if $m(\overline{AB}) = k \cdot m(\overline{AE})$ and $m(\overline{AC}) = k \cdot m(\overline{AF})$, where k is some positive real number, then $\overline{EF}$ is parallel to $\overline{BC}$.

8.4   SIMILAR FIGURES

In an earlier chapter you looked quite carefully at the concept of congruence and established the relationship of congruence as meaning same size and shape. Another relationship that figures sometimes have is the relationship of *similarity*. Two figures are similar if they are the same shape but not necessarily the same size.

PROBLEM 1   **a.** Draw two squares of different size on a sheet of paper. Locate them randomly on the sheet.

b. Describe a sequence of mappings that will map one of the squares onto the other. **Hint:** Consider both dilations and rigid motions.

c. If two figures are similar, do you think you could always map one to the other? Give a counter-example if necessary.

HOMEWORK EXERCISES

For each pair of similar figures in the following exercises, describe a composition of mappings (dilations, rigid motions) that will map one figure onto the other.

1.

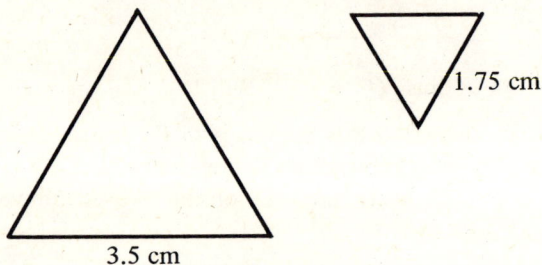

3.5 cm

1.75 cm

FIGURE 8.5

2.

Figure 1 = Square $ABCD$
Figure 2 = Square $EFGH$
$m(\overline{AB})$ = 4 cm
$m(\overline{EF})$ = 3 cm

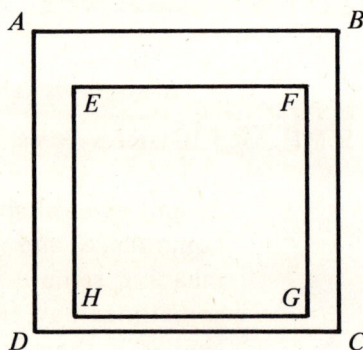

FIGURE 8.6

The preceding problem and homework exercises should have suggested a relationship between similarity of two figures and the existence of a composition of mappings to map one figure to the other. You can use this relationship to write a more precise definition of similarity.

DEFINITION: Two sets of points, *X* and *Y*, are *similar* (denoted *X* ~ *Y*) if and only if there exist a dilation, *D*, and rigid motion, *M*, such that

$$X \xrightarrow{D \circ M} Y$$

Note: Either *D* or *M* could be an identity mapping.

PROBLEM 2   For each of the following figures, decide whether or not set *X* is similar to set *Y*. If they are similar, specify precisely the dilation and rigid motion needed to map set *X* to set *Y*.

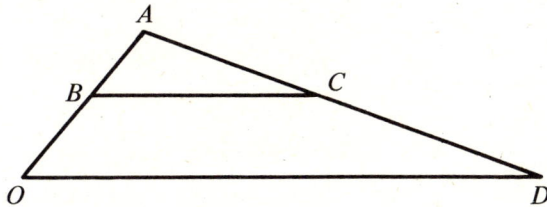

FIGURE 8.7

a.   Given: $\overline{BC} \parallel \overline{OD}$

$$m(\overline{AB}) = 2$$
$$m(\overline{AO}) = 5$$

Set $X = \triangle ABC$
Set $X = \triangle AOD$

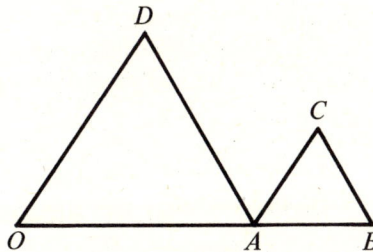

FIGURE 8.8

b.   Given: $m(\overline{AB}) = \frac{1}{3} m(\overline{OA})$

$$A \in \overline{OB}$$

$$\measuredangle DAO \cong \measuredangle CAB$$
$$\measuredangle DOA \cong \measuredangle CBA$$

Set $X = \triangle DOA$
Set $Y = \triangle CBA$

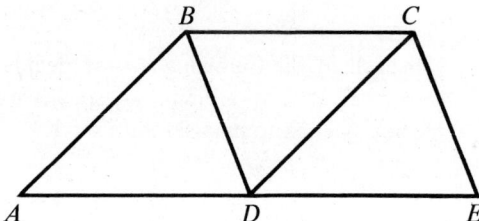

FIGURE 8.9

c. Given: $m(\measuredangle BAD) = 45°$

$$\overline{AB} \cong \overline{BC} \cong \overline{CD} \cong \overline{DA}$$

$$\overline{BD} \parallel \overline{CE}$$

Set $X = \triangle ABD$
Set $Y = \triangle DCE$

d. Use the same diagram and given conditions as part c.

Set X = quad $ABCD$
Set Y = quad $ECBD$

PROBLEM 3 Decide if the following statements are true or false. Provide counterexamples for those that are false. Justify those that are true.
a. Congruent figures are similar.
b. Similar figures are congruent.
★c. Similarity is an equivalence relation on the set of geometric figures.

8.5 AREA AND VOLUME OF SIMILAR FIGURES

PROBLEM 1 a. Let A, B, C be points in a plane with coordinates (1,1), (4,1), and (1,5) respectively.
1. Draw a set of axes and scale them in centimeters.
2. Locate A, B, and C on your graph.

292

3. $m(\overline{AB}) =$

   $m(\overline{AC}) =$ area $\triangle ABC =$

   $m(\overline{BC}) =$ perimeter $\triangle ABC =$

4. Consider a pyramid with $\triangle ABC$ as base and a height of 7 cm. What is the volume of the pyramid?

b. 1. Find the images of A, B, and C for $D_{(0,0),2}$. Label them A', B', and C'.

2. $m(\overline{A'B'}) =$

    $m(\overline{A'C'}) =$

    $m(\overline{B'C'}) =$

    area $\triangle A'B'C' =$

    perimeter $\triangle A'B'C' =$

3. What is the height of a pyramid which is similar to the pyramid in part **a** which has $\triangle A'B'C'$ as a base? Find the volume of the pyramid.

c. Repeat the procedure by mapping A', B', and C', with $D_{(0,0),\frac{3}{2}}$. Plot and label the image points A'', B'', and C''.

Find the comparative measures.

d. $\triangle ABC \overset{D_{(0,0),2}}{\rightarrow} \triangle A'B'C'$

    $\triangle A'B'C' \overset{D_{(0,0),\frac{3}{2}}}{\rightarrow} \triangle A''B''C''$

What dilation takes $\triangle ABC \rightarrow \triangle A''B''C''$?

e. Examine and compare the results obtained for parts **a** through **d** of this problem. Does there appear to be a relationship between lengths, areas, and volumes of similar figures? What is it? **Hint:** Compare scale factors and comparative measures.

f. Summarize your findings by completing the following statements.

If $\overline{AB} \overset{D_{P,k}}{\rightarrow} \overline{A'B'}$, then $m(\overline{A'B'}) =$ _____ $\cdot m(\overline{AB})$.

If plane region $R \xrightarrow{D_{P,k}}$ plane region R', then the area of $R' = $ _____.

If solid region S' is the image of solid region S for a dilation $D_{P,k}$, then _____.

8.6 USING WHAT YOU KNOW ABOUT SIMILARITY IN PROBLEM SOLVING

PROBLEM 1   Two photographs from the same negative are such that the area of one is 25 times the area of the other. If the smaller photograph is 1½ cm by 2 cm, what are the dimensions of the larger photograph?

PROBLEM 2   A man calculated that he needed 72 square tiles, 30 cm on each side, to tile his kitchen floor. When he went to buy them, he saw some tiles that were smaller squares, 15 cm on a side, that he liked better. He bought 144 of the smaller tiles. Explain why he found his purchase unsatisfactory when he tried to tile his kitchen floor.

PROBLEM 3   If two similar cylindrical jars hold ½ liter and 4 liters respectively, and the height of the smaller jar is 15 cm, find the height of the larger jar.

PROBLEM 4   A 50-cm long scale model of a swimming pool holds 9 liters of water. If the real pool is 25 m long, how much water is required to fill it?

PROBLEM 5   What are the missing elements of the set $\left\{3\frac{1}{2}, \text{—}, \text{—}\right\}$ if its elements are proportional to those of the set $\{1, 4, 5\}$? Describe a mapping that maps the first set to the second.

PROBLEM 6   It has been observed that during a total solar eclipse the moon passes between an earthbased observer and the sun so as to just block the sun completely. This can be represented schematically as shown in Figure 8.10.

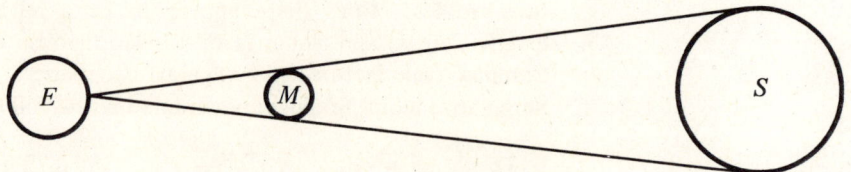

FIGURE 8.10

It is known that the earth-sun distance is almost 400 times the earth-moon distance. The earth-moon distance is about 38×10^4 km. What is the earth-sun distance?

If the diameter of the moon is about 3500 km, what is the diameter of the sun?

What is the ratio of the sun's volume to that of the moon?

These methods were used by the Greek astronomer Aristarchus around 250 BC. His results were poor because none of the distances involved were known accurately. However, the methods used were still applicable when better estimates of the distances were made.

PROBLEM 7 An x-ray photograph always shows an image which is larger than the object under study, as can be seen from Figure 8.11. The bone (or other organ) is a distance x away from the plate because of the flesh that surrounds it. Similarly, the source of the x-rays is a distance y from the bone.

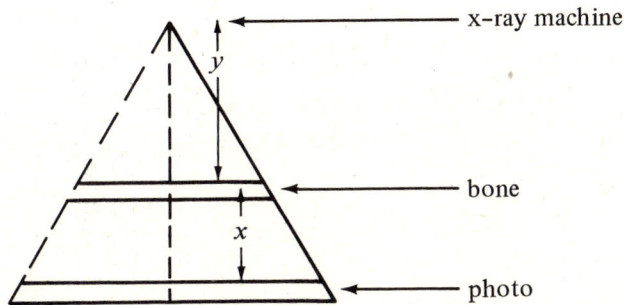

FIGURE 8.11

In terms of x and y, what is the ratio of the length of the bone to the length of its image?

If $x = 3$ cm, $y = 12$ cm, and the photograph shows an image 6 cm long, what is the actual length of the object photographed?

S U M M A R Y

In this chapter you have seen how measure can be incorporated into the definition of a mapping called a dilation. Dilations share some properties with certain types of rigid motions, but do not generally map figures to congruent figures. Figures are always mapped to similar figures by a dilation. In the next chapter, you move away from geometric mappings and look at an important area of mathematics that has a great deal of impact on our life—probability and statistics.

Probability and Statistics

9

9.1 INTRODUCTION

In your progress through the preceding chapters, you have studied the structuring of sets by operations such as $(Z_7, +)$ and then by relations, $(Q, >)$. You looked at sets structured by mappings of various kinds in Chapter 5 and mappings involving geometric figures such as rigid motions, measurement, and dilation in Chapters 6, 7, and 8.

This chapter will look at some special mappings that you are likely to encounter in the future and some special properties of some of these mappings.

The study of congruence and similarity focused on geometric mappings of points to points and figures to figures. Binary operations mapped ordered pairs of elements to a single element $((3,4) \overset{+}{\to} 7$, etc.) and measurement mapped figures to numbers $(\overline{AB} \to m(\overline{AB}))$. It is possible to map any kind of element to any kind of element providing you conform to the definition of mapping: each domain element is mapped to a unique (exactly one) range element. Most of the mappings in this chapter will map a set of numbers to a single number.

9.2 STATISTICS

You will recall that in Section 3.1 you investigated a binary opera-tion, γ, which mapped ordered pair (a,b) to the number $\dfrac{a+b}{2}$.

This operation assigns the *average* of 2 numbers to the ordered pair of numbers. If this operation is generalized to a mapping that assigns the *mean* (average) of any set of numbers to that set, the result will be one example of a *statistic*.

A statistic is a number assigned to a set of numbers which summarizes or indicates some characteristic or property of that set of numbers.

Example: Let M be the mapping which assigns the mean to a set of numbers. Then $\{3, 2, 5, 6\}\ \underline{M}\ 4$, $\{17.2,\ ^-6.1, 32.5,\ ^-0.6\}$ $\underline{M}\ 10.885$.

Thus, the mean of a set of numbers is a statistic for that set of numbers.

PROBLEM 1   a. Find the mean image for each set of numbers for mapping M.
1. $\{28, 35, 42, 50\}$
2. $\{^-2, 0, 3,\ ^-8, 20\}$
3. $\{10, 35, 60, 70, 96\}$

b. Write an algebraic formula to define the mean for a set of n num-bers, $\{x_1, x_2, \ldots, x_n\}$. Mean of $\{x_1, x_2, \ldots, x_n\}$ = _____.

c. Find an example of a set whose mean is an element of the set.

d. Find an example of a set whose mean is equal to the difference of the largest and smallest elements in the set.

e. Is the mean of a set always between the largest and smallest ele-ment of the set?

PROBLEM 2   Five teachers who teach the same course gave a standardized achieve-ment test to their students at the end of the year. The means for the scores of the five classes were all the same, 75. What would you con-clude about the classes? About the teachers? Why?

PROBLEM 3   Suppose the scores for the 5 classes in Problem 2 are:

Class 1: $\{74, 74, 75, 75, 75, 75, 75, 75, 76, 76\}$

Class 2: $\{50, 50, 50, 50, 50, 100, 100, 100, 100, 100\}$

Class 3: $\{71, 72, 73, 74, 75, 75, 76, 77, 78, 79\}$

298

Class 4: $\{50, 65, 68, 70, 75, 80, 82, 85, 100\}$

Class 5: $\{0, 0, 73, 77, 100, 100, 100, 100, 100, 100\}$

Would you change any of the conclusions you reached in Problem 2? Which ones? Why?

The data in Problem 3 indicate the inadequacy of the mean by itself to describe a set of numbers.

PROBLEM 4 Suggest and/or invent some statistics that might be useful to describe a set of numbers. Define them algebraically if possible, and compute the value of each statistic for each of the sets of scores in Problem 3. How does each statistic help describe the set of scores?

You found several statistics in Problem 4 that describe a set of numbers. Three statistics—*median, mode,* and *range*—are commonly used in addition to mean. These statistics are numbers assigned by mapping to a set S as follows:

The *median* of S is a number d for which half of the numbers in S are above d and half are below d.

The *mode* of S is the number in S which occurs most often (if such a number exists).

The *range* of S is the difference between the maximum number in S and the minimum number in S.

For the set $S = \{1, 3, 4, 4, 5, 6, 6, 6, 8\}$ the median is 5, the mode is 6, and the range is 7.

PROBLEM 5 Check to see if any of the statistics you derived in Problem 4 are the same as these three statistics.

HOMEWORK EXERCISES

1. Find the median, mode, and range for each set of scores in Problem 3.

2. How do the additional statistics computed in Exercise 1 help describe the nature of the sets of scores in Problem 3? Compare mean, median, mode, and range for the 5 classes.

3. Consider the following 3 sets of numbers.

$$S_1 = \{1, 2, 2, 10, 10, 10, 18, 18, 19\}$$
$$S_2 = \{1, 9, 9, 10, 10, 10, 11, 11, 19\}$$
$$S_3 = \{1, 5, 6, 10, 10, 10, 14, 15, 19\}$$

a. Would you say the sets of scores are alike? How?
b. Are the sets of numbers different? How?
c. Find the mean, median, mode, and range for each set.

The characteristic that is different for the sets of numbers in previous Homework Exercise 3 is the dispersion or spread of the numbers around the mean. The set, S_1, has an equal number of numbers in the middle and at the ends with none between; S_2 has only 1 number at each end with most clustered around the middle; S_3 has a more uniform dispersion of the numbers. What is needed, then, is a statistic that indicates the dispersion or spread of the numbers in a set about the mean of that set.

PROBLEM 6   Complete Table 9.1 and construct similar tables for S_2 and S_3 in exercise 3 of the previous homework.

TABLE 9.1 $S_1 = \{1, 2, 2, 10, 10, 10, 18, 18, 19\}$

Mean (*m*)

Element	Value x	Deviation from Mean $x - m$	Squared Deviation from Mean $(x - m)^2$
1	1	⁻9	81
2	2	⁻8	64
3	2		
4			
5			
6			
7			
8			
9			
Totals			

Sum of Scores =   Sum of Deviations (*sd*) =

Sum of Squared Deviations $(sd^2) =$

Mean Deviation $(md) = \dfrac{sd}{n} =$

Mean Squared Deviation $(md^2) = \dfrac{sd^2}{n} =$

Compare the three sets, S_1, S_2, and S_3, and the completed tables for each, and fill in Table 9.2.

TABLE 9.2

	Mean	*Median*	*Mode*	*Range*	*sd*	*sd²*	*md*	*md²*
S_1								
S_2								
S_3								

Which of the statistics in Table 9.2 would be the best for comparing dispersion about the mean for sets of numbers? Explain your choice.

The mean squared deviation of a set of numbers is a universally accepted statistic for measuring dispersion about the mean. It is called *variance* (hereafter denoted *v*) of the set. The square root of the variance is called the *standard deviation (s)*. Since the size of *s* is directly related to the size of *v*, *s* is also a measure of dispersion about the mean. You will note some interesting properties of *s* later.

HOMEWORK EXERCISES

4. Compute the standard deviation for each set of scores in Problem 3.

5. **a.** Flow chart and write a program in BASIC to compute and print out the mean for any set of 10 numbers.
   b. Modify your flowchart and program so it will work for any number of numbers.
   c. Again modify your flowchart and program so it will also compute and print out the standard deviation for the set of numbers.

9.3 FREQUENCY DISTRIBUTIONS

A common way to display data such as that of Problem 3 and Homework Exercise 3 is to make a graph called a *frequency histogram* (bar graph). A frequency histogram is constructed by plotting all possible values of some variable with the frequency for each value and then constructing "bars" for each nonzero frequency as shown in Figure 9.1.

Example:

Let $S = \{1, 2, 2, 3, 4, 4, 4, 5, 5, 6, 7, 8, 9, 9, 10\}$

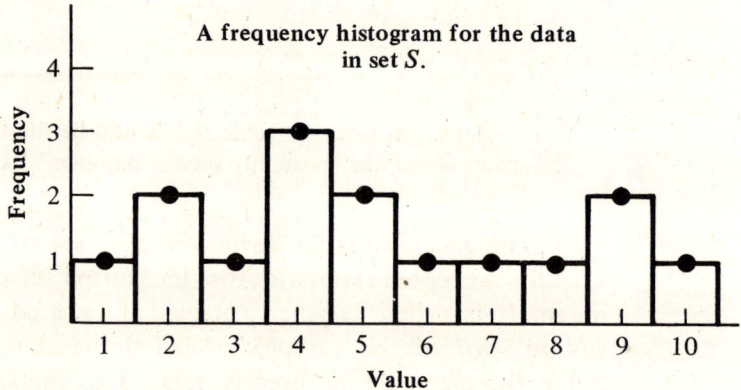

A frequency histogram for the data in set S.

FIGURE 9.1

PROBLEM 1   Toss 3 coins and record the number of heads obtained. Repeat the experiment 20 times and summarize the data in the following frequency table.

TABLE 9.3

Number of heads	Frequency
0	
1	
2	
3	

Make a frequency histogram for the data obtained.

The purpose of frequency histograms and the variations that follow is to "picture" the data in such a way that some of the characteristics are obvious.

A *frequency polygon* (Figure 9.2) is constructed in the same way as a histogram except the plotted points are connected with line segments instead of bars.

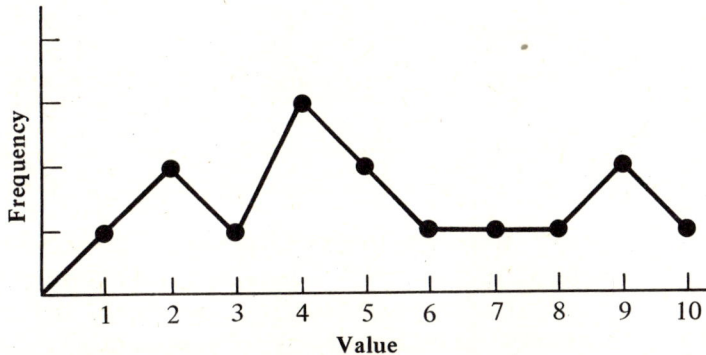

FIGURE 9.2

If you smooth out the frequency polygon, it becomes a *frequency curve*. Frequency curves are most appropriate when both the number of possible values and the range of frequencies are large.

HOMEWORK EXERCISES

1. Construct a frequency histogram, frequency polygon, and frequency curve for the data given in Table 9.4.

2. Find the mean for the set of data described in Exercise 1.

★3. Write an algebraic formula for computing the mean for a set of scores when given the scores and their frequency instead of the actual list of all scores. Let $x_1, x_2, \ldots, x_n$ represent the different scores and $f_1, f_2, \ldots, f_n$ represent the frequencies of each score, respectively.

TABLE 9.4

Number	Frequency
0	0
1	2
2	5
3	3
4	4
5	7
6	10
7	12
8	5
9	3
10	1

Some examples of frequency curves for some sets of data are shown in Figure 9.3. The dotted lines represent the mean score, *m*.

Frequency histograms, polygons, and curves as well as frequency tables are all ways of representing a *frequency distribution* or simply a *distribution*.

The curves in parts **a** and **b** of Figure 9.3 represent skewed distributions; part **c** shows a *normal* distribution, part **d** shows a bimodal distribution, and part **e** shows a uniform distribution.

PROBLEM 2   Try to think of real data that you would expect to have a skewed distribution, a normal distribution, a bimodal distribution, and a uniform distribution. List your guesses.

The distribution of most interest and the one you are most likely to see or hear about in the future is the normal distribution. Some important characteristics of a normal distribution are listed below.

1. Approximately 68 percent of the numbers of a normal distribution fall within 1 standard deviation of the mean (34 percent on each side).

2. Approximately 96 percent of the numbers of a normal distribution fall within 2 standard deviations of the mean (48 percent on each side).

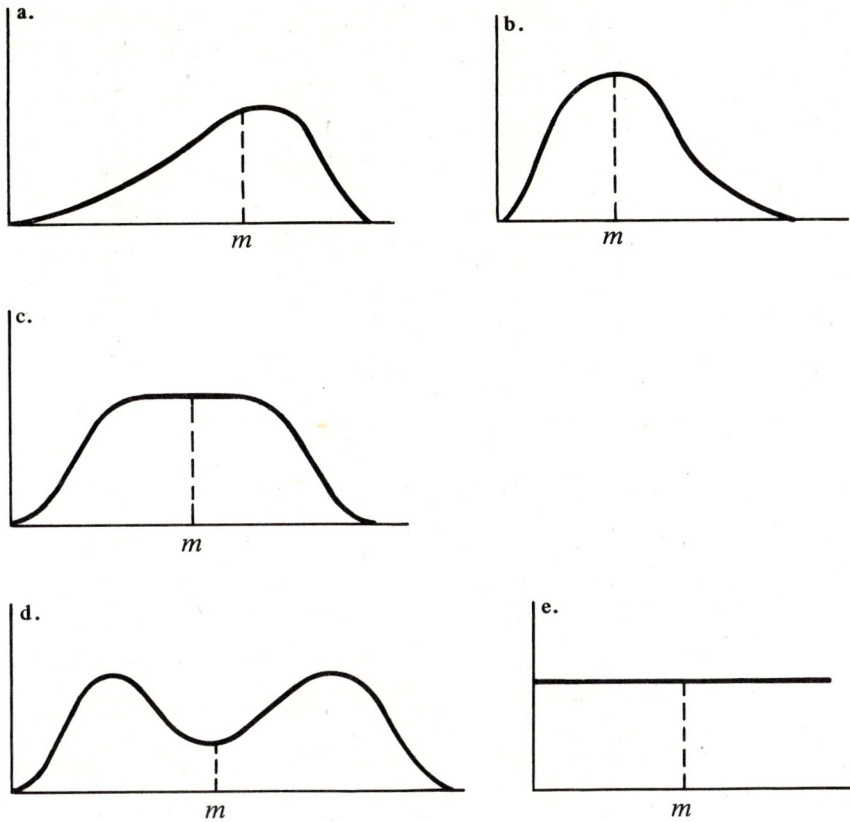

FIGURE 9.3

Example:

The frequency curve in Figure 9.4 shows a normal distribution with a mean of 50 and a standard deviation of 10.

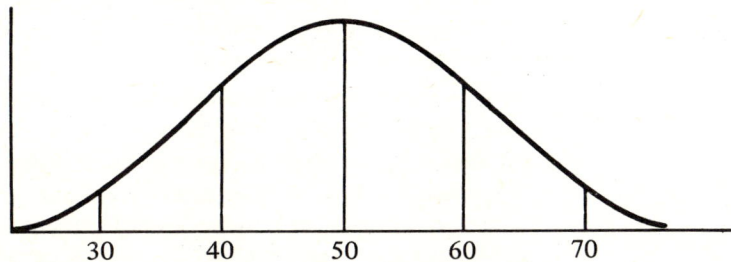

FIGURE 9.4

Since the distribution is normal, we would expect 68 percent of the numbers to fall between 40 and 60, and 96 percent of the numbers to fall between 30 and 70.

PROBLEM 3   Suppose a distribution of heights (measures in centimeters) of 500 males is normal with a mean of 173 and a standard deviation of 8. Sketch a graph of the distribution described indicating the mean and scores 1 and 2 standard deviations on either side of the mean. How many of the heights would be expected:

  a. between 165 and 181?
  b. between 157 and 189?
  c. between 157 and 173?
  d. less than 157?
  e. greater than 157?
  f. greater than 181?

HOMEWORK EXERCISES

4. What kind of distribution would you expect to obtain for each of the following sets of data?

  a. heights of 18-year-old persons in the United States
  b. numbers formed by the last 2 digits of each persons's social security number
  c. IQ scores for adults in the United States
  d. grade point averages for college freshmen in the United States

★5. **a.** What kind of distribution do the data in Homework Exercise 1 in this section have?
  b. Find the mean and standard deviation for that data. You might want to use a calculator.
  c. Calculate the percent of scores in each of the following ranges (m denotes the mean, and s denotes the standard deviation).
    1. $m - s$ to $m + s$
    2. $m - 2s$ to $m + 2s$
  Do your results support your answer to part **a**? Why?

6. **a.** Read the instructions for the experiment in part **b** that follows. Before carrying out the procedure, guess what kind of distribution you will obtain for frequencies of the numbers.
  b. Experiment: Take 4 equal size pieces of paper and number them 1 through 4. Place the pieces of paper in a paper bag and shake them up. Reach in and without looking pick out one piece of paper. Record

the number on the paper. Replace the paper in the bag. Repeat the shake, pick, record, and replace sequence 20 times. Summarize your results in Table 9.5.

TABLE 9.5

Number	Frequency
1	
2	
3	
4	

c. Make a frequency polygon for the data.
d. What kind of distribution did you obtain? Is it what you expected in part **a**?

In question **6a** of the previous homework exercise, you may have guessed that you would obtain a uniform distribution since each number is just as likely to be picked as any other. Theoretically, the experiment will yield a uniform distribution in the long run. As the number of pieces of paper picked increases, the distribution becomes closer to being truly uniform. You will look at this concept again later in the chapter.

9.4 PROBABILITY

Before proceeding with the discussion of statistics, you will take a brief look at the concept of probability. You have undoubtedly encountered the word or concept of probability in many ways.

PROBLEM 1   What does each of the following statements mean?
a. The probability of getting a head when you toss a coin is ½.
b. There is a 20-percent chance of precipitation today.
c. The odds in favor of the Yankees winning the pennant are 3 to 7.

Probability is a measure of the likelihood that a given event will happen. This measure can be stated in terms of a number, a percent, or odds, as indicated by the statements in Problem 1.

Some important terms that will be needed in the development of probability are outcome, sample space, and event.

An *outcome* is an observable result of some experiment or action. Tossing a head, observing rain, and the Yankees' winning the pennant are examples of outcomes. When a 6-faced die is tossed, there are 6 possible outcomes: 1, 2, 3, 4, 5, & 6. A *sample space* is the set of all possible outcomes, $\{1, 2, 3, 4, 5, 6\}$ when tossing a die; $\{$Yankees win, Redsox win, ... $\}$ for the pennant race; $\{$rain, snow, sleet, hail, clear, cloudy $\}$ for observing weather. An *event* is a set of one or more outcomes from the sample space. $\{$Redsox win$\}$ is an event. Precipitation = $\{$rain, sleet, snow, hail $\}$ is a weather event. Tossing an even number = $\{2, 4, 6\}$ is an event when tossing a die.

PROBLEM 2   a.   List the sample space for tossing:

1 coin

2 coins

3 coins

b.   For the second and third sample spaces you listed, list the outcomes from the sample space that would belong to the event: Getting two heads. Did you consider the outcome of a head on the first coin and a tail on the second coin different from a tail on the first coin and a head on the second? Are they the same physical result?

c.   List the possible outcomes in the sample space when tossing two dice (6-faced). Represent each outcome as *x-y* where *x* is the number on the first die and *y* is the number on the second die. **Note:** Consider 2-3 different from 3-2.

    1.   How many outcomes are there?

    2.   List the outcomes in each of the indicated events.

E_2: The sum of the faces = 2

E_3: The sum of the faces = 3

    ·   ·

    ·   ·

    ·   ·

E_{12}: The sum of the faces = 12

Define the *probability* of an event, *E*, written $P(E)$ as:

$$P(E) = \frac{\text{number of outcomes in } E}{\text{number of outcomes in the sample space}}$$

Example:

From Problem 2c parts 1 and 2:

$$P(E_2) = \frac{1}{36}$$

$$P(E_3) = \frac{2}{36}$$

PROBLEM 3 a. Using the definition of probability, compute the probabilities of each event listed in **Problem 2c**, part 2:

  b. Which sum is most likely to occur? Least likely?

  c. What does E represent if $E = E_2 \cup E_3 \cup \ldots \cup E_{12}$? Find $P(E)$.

  d. Let A be the event: the sum of the faces is even. Find $P(A)$. Use the results of parts **c** and **d** to find the probability of tossing an odd sum.

  e. Make a frequency table and histogram for the expected frequency of each sum in 36 tosses of the dice. Does the distribution appear normal?

  f. Actually toss a pair of dice 36 times recording the sum obtained for each toss. Compare your results with the expected results. Can you draw any conclusions about the fairness of the dice you used? What are your conclusions?

PROBLEM 4 Complete Table 9.6.

TABLE 9.6

Number of coins	1	2	3	4		n
Number of outcomes having exactly 2 heads	0	1				
Total number of outcomes	2	4				
Probability of tossing 2 heads	$\frac{0}{2} = 0$	$\frac{1}{4}$				

HOMEWORK EXERCISES

1. List the outcomes that yield exactly one head when tossing 1 coin, 2 coins, 3 coins.

2. Make a table like Table 9.6, and find the probability of getting 1 head when tossing *n* coins.

3. Suppose you are in a bingo game where there are cubes numbered from 1 to 15 for each letter in the word *bingo*.
   a. How many cubes are there?
   b. What is the probability that G-10 will be drawn on the first draw?
   ★c. Suppose there have been 20 cubes drawn (and not replaced in the drawing bin) and you need B-2 to win. What is the probability of your winning on the next draw? On the draw after the next?

4. Fill in the first four lines of Figure 9.5. Much of the information can be found in Problem 4 and Homework Exercises 1, 2. When you have completed the first 4 lines, seek out patterns that might aid you in completing lines 5 and 6 without going back to consideration of coins. The number in parentheses indicates the number of heads. The number above the line indicates the number of outcomes with that number of heads.

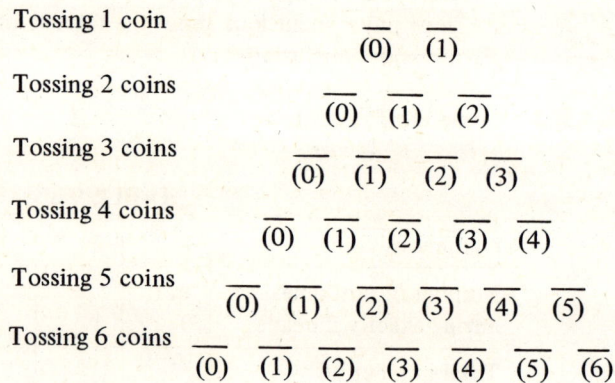

Tossing 1 coin
$$\overline{\quad}\ \ \overline{\quad}$$
$$(0)\ \ (1)$$

Tossing 2 coins
$$\overline{\quad}\ \ \overline{\quad}\ \ \overline{\quad}$$
$$(0)\ \ (1)\ \ (2)$$

Tossing 3 coins
$$\overline{\quad}\ \ \overline{\quad}\ \ \overline{\quad}\ \ \overline{\quad}$$
$$(0)\ \ (1)\ \ (2)\ \ (3)$$

Tossing 4 coins
$$\overline{\quad}\ \ \overline{\quad}\ \ \overline{\quad}\ \ \overline{\quad}\ \ \overline{\quad}$$
$$(0)\ \ (1)\ \ (2)\ \ (3)\ \ (4)$$

Tossing 5 coins
$$\overline{\quad}\ \ \overline{\quad}\ \ \overline{\quad}\ \ \overline{\quad}\ \ \overline{\quad}\ \ \overline{\quad}$$
$$(0)\ \ (1)\ \ (2)\ \ (3)\ \ (4)\ \ (5)$$

Tossing 6 coins
$$\overline{\quad}\ \ \overline{\quad}\ \ \overline{\quad}\ \ \overline{\quad}\ \ \overline{\quad}\ \ \overline{\quad}\ \ \overline{\quad}$$
$$(0)\ \ (1)\ \ (2)\ \ (3)\ \ (4)\ \ (5)\ \ (6)$$

FIGURE 9.5

Unfortunately, the definition for probability doesn't really cover the weather situation or the odds in favor of the Yankees winning the pennant. The definition covers only the situation where you are concerned with *random* events of a finite sample space.

If the probability of precipitation was based only on the listing of possible weather outcomes given as an example for the term *outcome,* then the probability of precipitation would be 4/6 or $66^2/_3$ percent all the time. The interpretation for the probability (or chance) statement for weather is that "under similar atmospheric conditions in the past it has precipitated 20 percent of the time."

Similarly, in the case of the odds for the Yankees winning the pennant, a direct application of the definition is inappropriate. Since there are twelve teams in the American League, the probability of the Yankees winning the pennant would be 1/12 if the winner were picked at random. However, the winner is not picked at random so the probability is not 1/12. When the odds-makers make statements like "The odds in favor of the Yankees' winning the pennant are 3 to 7," they are giving the Yankees 3 chances in 10 to win, based on the odds-makers' guess of the Yankees' ability relative to the abilities of the other teams in the league. They are guessing that the probability of the Yankees winning is 3/10. (*Odds* state the ratio of favorable to unfavorable outcomes; thus 3 to 7 implies 3 out of a possible 10 are favorable, which implies a probability of 3/10.)

Often, probability statements are based on hunches and past performance when no computational scheme is available. However, probability statements consistently place the measurement of the likelihood of an event's happening between 0 and 1, either as a number, or a percent between 0 and 100 (which can be converted to a number between 0 and 1), or as odds x to y, which can be converted to a probability of $\dfrac{x}{x+y}$ which is a number between 0 and 1. Thus, probability is a mapping of sets to the number interval 0 to 1 (inclusive).

HOMEWORK EXERCISES

5. Odds are stated in favor of an event occurring. For example, the odds for having rain today are 2 to 3. However, odds can be given for the event not happening. Assuming the odds for rain are true, what are the odds against having rain today? Find the odds in favor and the odds against each event in Problem 2c, part 2 of this section.

6. Complete Table 9.7.

TABLE 9.7

Probability of event	Percent chance the event will happen	Odds in favor of the event happening
$\dfrac{1}{7}$		
	25%	
.3		
		1 to 6
	14.3%	
		4 to 5

7. If S denotes a sample space, and E is some event of the sample space, then we know $E \subseteq S$. The complement of E, denoted $\overline{E}$, is equal to the set of all outcomes in S that are not in E.

What is the probability of $\overline{E}$ relative to $P(E)$?

$$P(\overline{E}) =$$

Find the complement and probability of the complement for each event in Problem 2c part 2.

8. Find the probability, percent chance in favor, and odds in favor of the complement occurring for each event in Table 9.7.

★9. Flow chart a procedure and write a program in BASIC to calculate the odds against, percent chance in favor, and probability when you input the odds in favor for the event.

9.5 PREDICTION

Now look at some specific examples of how probability and statistics are used for prediction. The important kinds of prediction you will look at, using specific examples, are: predicting future events from knowledge of past events (Probability and Life Insurance) and predicting the value of a statistic for a set of data from knowledge of the value of that statistic for a subset of the data (Sampling a Population).

Probability and Life Insurance

On the basis of past experience, a life insurance company can make predictions as to how many policyholders will be expected to die each year and will thus know how much money will have to be paid out in claims. Although the life insurance company never knows who will die, it can predict how many people in a particular age group will die during one year.

Life insurance companies compile *mortality tables* based on their experience with policyholders in the past. To make up a mortality table, the life insurance actuary must determine the death rate for each age group insured.

PROBLEM 1 Here is the mortality experience of an insurance company, Company A, with its group of 16-year-old policyholders.

a. The number of policyholders (age 16) at the start of the year was 5844. Nine died during the year. So the probability of dying, $P(D)$, was $\dfrac{9}{5844}$.

Life insurance companies state probability of dying as a death rate per 1000. If we let x = death rate per 1000, then $x = 1000 \cdot P(D)$.

Compute the death rate per 1000 for the 16-year-olds insured by Company A.

b. At age 17, there were 5835 of the previously insured 16-year-old policyholders living at the start of the year. The company added 263 new 17-year-old policyholders for a total of 6098. During the year, 10 died. Complete Table 9.8.

TABLE 9.8

Age	Number of old Policyholders	Number of new Policyholders	Total number Policyholders	Number Dying	Death Rate
16		5844	5844	9	
17	5835	263		10	
18		345		11	

c. Once the death rate per 1000 for each age is known, you can construct a mortality table for any group of people at successive years of their lives. Complete the following mortality table.

TABLE 9.9

Age	Number Policyholders at start of year	Death Rate per 1000	Predicted Number of Deaths during year
16	50,000	1.54	77
17		1.64	
18		1.71	

Table 9.9 is just an abbreviated table for a single base number (50,000). Mortality tables are used to compute the necessary premium a company must charge to insure a policyholder.

When a person buys an insurance policy, he or she agrees to pay the insurance company a certain amount of money (premium), usually on an annual basis. This amount can be fixed or vary from year to year depending on the type of insurance bought.

PROBLEM 2   Suppose 50,000 16-year-olds purchase $1000 policies (policies that pay $1000 upon the death of the insured). Use the data in Table 9.9 for the following computations.

a. Find the predicted amount of money the company will pay in claims during the coming year, the second year, and the third year, assuming no new policyholders are added.

b. Suppose the company determines that they must add 15 percent of expected claims as administrative costs. Find the administrative costs for each year of the three-year period.

c. Using your results of parts a and b, compute the necessary premium the company must charge its policyholders each year during the three-year period.

This example, although over-simplified, shows the use of past experience, in the form of mortality statistics, to compute a probability of dying (or an equivalent death rate) for a given age group. This information is then used to predict the future claims of policyholders, which in turn is used to compute the necessary premiums. The data is updated regularly and computed independently by each company. This model can be generalized to automobile, personal injury, and homeowners insurance as well as other types of insurance.

PROBLEM 3   List the data an insurance company would need in order to compute premiums for automobile liability insurance for the 16 to 20 age group.

HOMEWORK EXERCISE

1. Suppose 10,000 persons in the 16 to 20 age group buy automobile liability insurance from Company Q. Past experience for Company Q shows the probability of a policyholder in this age group having an accident involving a liability claim payable by the company is 0.25. The average claim paid by the company is $800. The company calculates its administrative costs at 20 percent of claims paid. Compute the annual premium the company must charge each policyholder in this age group.

Sampling a Population

The prediction of population values of a statistic from sample (subset of the population) values of the statistic is a major role of statisticians. Opinion polls, quality control, educational experiments, and biomedical research are all concerned with the estimation (prediction) of population statistics from sample statistics. For example, opinion pollsters usually sample less than 2000 persons and estimate opinions for the whole country based on the data collected from the sample. Product manufacturers sample as few as 5 or 6 products from a production line to predict the overall quality of the entire production run. Medical researchers will test new drugs or procedures on a sample of individuals to predict the effects with more general use.

The usefulness of such predictions is entirely dependent upon the accuracy of the prediction. Thus, one must determine the most accurate procedure for predicting a population statistic and the level of accuracy to expect using that procedure. The accuracy of prediction using sample statistics can be determined by looking at the characteristics (distribution) of sample statistics when many samples are taken from a single population.

PROBLEM 4   Experiment to predict the average height and weight of the students in the class using only a sample of the data available.

Class Preparation:

Each student writes his/her height (nearest centimeter) and weight (nearest gram) on identical slips of paper. All slips are placed in a shoe box that is then sealed closed. The box should be shaken vigorously for at least 15 seconds. Cut a hole in one end of the box just large enough for a hand to reach in.

Procedures for Each Student:

Select 5 slips from different parts of the box and record the height and weight written on each slip. Replace the slips in the box and again shake vigorously.

PROBLEM 5   a. Find the mean, median, range, and variance for the heights of your sample. Record your results in column 1 of Table 9.10.

TABLE 9.10

Statistic	Sample Value	Predicted Population Value	Actual Population Value
Mean			
Median			
Range			
Variance			

b. Based on the statistics of your sample, predict the mean, median, range, and variance of the heights of the total population (class). Record your predictions in column 2 of Table 9.10.

c. If the predicted population value differs from your sample value, explain your reasons for predicting the value you did.

d. Obtain the actual population values for each statistic and record the data in column 3 of Table 9.10. Examine the data in your table and in the tables of several other students. Based on your observations, conjecture relationships between sample values and population values for each statistic.

e. Suggest a procedure for testing your conjectures.

HOMEWORK EXERCISE

2. Make a table like Table 9.10. Carry out the procedures of Problem 1 for the weights obtained in the experiment. Compare your conclusions with the conclusions obtained in Problem 5.

PROBLEM 6 In the previous problem each student drew sample data on heights and weights from 5 students of the total class.

a. Did your sample mean equal the population mean for either height or weight? Check to see if any student obtained a sample mean equal to the population mean.

b. Obtain a listing of each student's sample means for height and weight. Make separate frequency curves showing the distribution of the set of sample means for heights and weights.

c. Find the mean of the sample means for heights and weights. Compare the mean of sample means to the appropriate population mean.

d. Conjecture what would happen to the distributions of sample means for heights and weights if we collected data from many more samples of size 5 from the population.

e. Obtain the variance for the set of sample means for heights and weights. Compare these with the respective population variances. Which is larger, the population variance or the variance of the set of sample means? Conjecture a relationship between the two variances.

f. Which do you think is the better predictor of population values: the sample mean or the sample variance?

★g. Ask your instructor if he/she has other similar data (population statistics and distributions of sample means) for you to examine. Test your conjectures about the variance of sets of sample means on the additional data.

The important idea developed in this section is that application of the concepts of probability and statistics can be used to predict future events or statistical values of a population from sample values. The importance of prediction to business, government, and education is obvious. With the recent growth of computer technology to minimize the tedium of the calculations involved, the usefulness and accuracy of prediction have improved greatly.

317

★9.6   THE CENTRAL LIMIT THEOREM

The previous section was intended to produce data suggestive of a very important relationship between sample statistics and population statistics. This relationship, known as the Central Limit Theorem (CLT), implies that as the number of elements, n, in samples from a population with mean, m_p, and variance, v_p, increases, the distribution of *sample means* approaches a normal distribution with mean, m_p, and variance, $v_{p/n}$.

Using the CLT you can predict a population mean from a sample and have some indication of the accuracy of your prediction. The following example will illustrate how this is done.

The Use of the CLT in Prediction

Suppose a sample of 10 students is chosen randomly from the student population of a large university. Each student is asked to give his/her current grade point average (GPA). The mean GPA for the sample is 2.5 and the variance of the sample is 0.4.

If you wish to predict the current GPA for the entire student population at the university, you can estimate the population variance with the sample variance giving us v_p = 0.4. The CLT indicates the variance of sample means is the population variance divided by the number in the samples. So, the variance of sample means for samples of 10 students is 0.4/10 = 0.04. If the variance is 0.04, then the standard deviation of sample means is 0.2, ($\sqrt{0.04}$ = 0.2).

The CLT also indicates that sample means are distributed *normally* about the population mean, m_p. Thus, the distribution of sample means for samples of 10 students would have a mean of m_p (unknown) and a standard deviation of 0.2 (assuming a population variance of 0.4). This distribution is pictured in Figure 9.6, page 319. The sample mean of 2.5 is from the distribution pictured in Figure 9.6.

PROBLEM 1   a.   What is the probability that the sample mean of 2.5 is within 0.2 of m_p? Within 0.4 of m_p?

b.   Is the sample mean GPA of 2.5 above or below the unknown population mean GPA? Can you tell?

c.   What is the probability that m_p is within 0.2 of 2.5? Within 0.4 of 2.5?

Are these questions the same as the questions in part **a**? Why?

d. If you predict that m_p is between 2.1 and 2.9, how certain are you of your prediction? (What is the probability that you are correct?)

e. If you predict that m_p is between 2.3 and 2.7, then the probability of being correct is _____.

Problem 1 shows that as the interval of prediction decreases, the probability of being correct also decreases. Since the probability of sample mean, m_{sample} being within two standard deviations of the population mean, m_p, is .96, you call the interval from $m_{sample} - 2s$ to $m_{sample} + 2s$ the 96 percent *confidence interval* for the population mean. Similarly $m_{sample} - s$ to $m_{sample} + s$ is the 68 percent confidence interval.

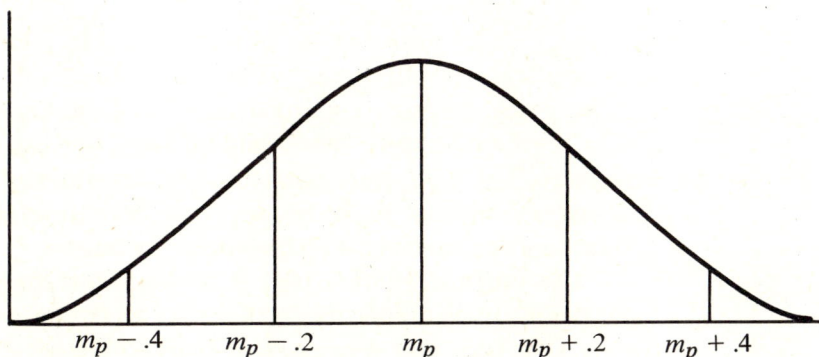

$$m_p - .4 \qquad m_p - .2 \qquad m_p \qquad m_p + .2 \qquad m_p + .4$$

FIGURE 9.6

HOMEWORK EXERCISE

Follow through the example and problem of this section using 40 instead of 10 as the number of students in the sample. Find 96 percent and 68 percent confidence intervals for the population mean. How does the increase in sample size affect the accuracy of prediction?

SUMMARY

Your exploration of mathematics has taken you through many topics. However, it is hoped that you have noticed the important underlying concepts that have been present throughout the material; namely, relations, mappings, and operations. Mappings, also known as functions in many books, are probably the most important concept of all of mathematics. Many specialized areas of mathematics are devoted to the study of special mappings and their uses—calculus, complex variables, and differential equations to name a few. The notion of structure is another important aspect which pervades much of mathematics. You looked at a great deal of structure in the course of your development. Groups and fields are examples of operational structures. Equivalence relations such as congruence and similarity are examples of relational structures. Induction and deduction also play a large role in any mathematical development. You used induction often when searching for patterns and relationships. Deduction was used to prove theorems about fields and groups, special relations, and mappings. The role of logical thinking was brought out in flow-charting as well as in those sections that developed deductive systems and/or careful investigations of structure.

The purpose of this text is to highlight these fundamental ingredients of mathematics in the hope that you would see some of the structure and processes of mathematics itself while focusing on structures within mathematics.

Index